RADIO
ROUND THE WORLD

The new B.B.C. transmitting station at Droitwich

RADIO
ROUND THE WORLD

by

A. W. HASLETT

Sometime Scholar of King's College, Cambridge

CAMBRIDGE

at the University Press

1934

CAMBRIDGE
UNIVERSITY PRESS

University Printing House, Cambridge CB2 8BS, United Kingdom

Published in the United States of America by Cambridge University Press, New York

Cambridge University Press is part of the University of Cambridge.

It furthers the University's mission by disseminating knowledge in the pursuit of education, learning and research at the highest international levels of excellence.

www.cambridge.org
Information on this title: www.cambridge.org/9781107418943

© Cambridge University Press 1934

First published 1934
First paperback edition 2014

A catalogue record for this publication is available from the British Library

ISBN 978-1-107-41894-3 Paperback

ACKNOWLEDGMENTS

Figs. 5 and 6 are reproduced in modified form from the *Proceedings of the Royal Society*, and Figs. 2, 8 and 9 from the *Proceedings of the Physical Society*. Plate 7 is taken, by permission of the Controller of H.M. Stationery Office, from *Applications of the Cathode Ray Oscillograph in Radio Research*. I am also indebted to the Council of Trinity College, Cambridge, for permission to reproduce a portrait of James Clerk Maxwell; to the British Broadcasting Corporation for a photograph of the new Droitwich transmitting station; to Sir Oliver Lodge for the loan of a portrait of Heinrich Hertz; to Dr Erwin Schliephake of Giessen University, Germany, for the two X-ray photographs reproduced as Plate 6; and to Professor Størmer of Oslo for permission to reproduce one of his many fine photographs of the Aurora Borealis. For much other personal assistance, generously given, I take this opportunity of expressing my thanks.

A. W. H.

CONTENTS

LIST OF ILLUSTRATIONS

RADIO ROUND THE WORLD

CHAPTER I

PRELUDE

THE why and the wherefore of wireless are in real danger of being lost in the flurry of achievement. The possessor of a good, but in no way unusual, receiving set can expect to be able to "tune in" during the quiet hours of the night to half a dozen or more American stations. The B.B.C. broadcasts regularly to the farthest corners of the Empire, and if reception is not perfect the voice of criticism is raised. It is as easy to talk to New York or Calcutta by radio-telephone as to one's next-door neighbour. Each year fresh triumphs are placed on record, and each year fresh catchwords are added to the jargon of radio—until, if pride did not prevent us, many of us would like to ask, quite simply, what is this wireless?

"Wireless", the expert replies, "consists of electro-magnetic vibrations in the ether, or (if you prefer it) in space-time." And we are left little wiser than before.

What, then, are electro-magnetic vibrations? How do they get into the ether? Why do our wireless sets give us, as a rule, only one out of the hundreds of stations which are broadcasting? Why do these vibrations in the ether (or space-time) sometimes fade out without warning as if they had never been? What are "atmospherics", and why are they so difficult to keep out?

The truth is that the waves of wireless—vibrations,

the expert called them—are just as genuinely waves as the rollers which beat in from the Atlantic upon the north coast of Cornwall or the ripples which a stone makes when it is dropped into a pond.

Even the longest rollers have a beginning somewhere, a centre of high wind, perhaps a thousand or more miles away. Where the ripples start from in a pond there is no doubt. These are the "transmitters" of water's more familiar waves. In each case energy is being put into the waves at the source—by the wind which is blowing harder out at sea, or by the person who lifts up the stone to drop it. In each case the waves follow one another at intervals which depend on their source.

So it is with wireless. The transmitter needs power, and the sort of waves which are produced depend on the way in which the power is applied. The power is supplied in the form of an electric current, and at whatever rate the engineer makes that current vibrate in his aerial, at that rate will successive waves follow one another out into space.

Then reception. We do not in the ordinary course have to bother about receiving water waves at all. We see them, or are knocked over by them. No outside aid is necessary. But a blind man sitting beside the Round Pond in Kensington Gardens could have his "receiver" if he wanted, "watching" on the bank to feel the ripples coming in. He could tie a piece of cotton to a cork, bobbing at the edge of the pond; and, with perhaps a little exercise of the imagination, feel the pull of the water as it sank between one wavelet and another.

That is the part of the wireless receiver—to move as

2

the wireless waves move, to vibrate in harmony as the string of a violin will "sing" when a second instrument sounds the note to which it is tuned.

Not all waves, it is true, behave in the same way. Already we have mentioned water waves, and light waves, and sound waves, as well as our original wireless waves. We would not know them by different names if they did not in many ways behave differently. But from these very differences we can learn much of what we want to know.

Ripples are sea rollers on a different scale, and the relation of wireless waves to light waves is the same. That will provide one line of attack. It will help us to understand why wireless goes round the world, and light does not; why even a mountain is not sufficient entirely to block out a wireless wave; and why wireless will penetrate, little diminished, the brick and mortar of our houses.

It is the purpose of this book to tell something of the story and to make clear—so far as it can be seen—what chapters may yet be written. The quest will take us as far as a hundred miles into the earth's atmosphere, as far north as the Arctic, and as far out into space as the sun itself. Wireless knows nothing of the limits of commerce. Like light it can pass through empty space. Its realm is the universe.

And, at the end, we shall see that radio is finding unexpected applications in medicine and as a means of weather forecasting. Nor have its powers been yet fully exploited either in warfare or in the promotion of safety at sea. If the past and the present of radio are full of interest, its future suggests tantalising possibilities.

I-2

CHAPTER II

FROM PROPHECY TO PROOF

I HAVE only one objection to wireless as a subject. Its story begins with a mathematician—a mathematician, it is true, who was at least half a scientist, and a mathematician, also, who was capable of being the life and soul of that mysterious body, the Red Lions, the unofficial dining club of the British Association, which takes its name from the Red Lion Inn at Birmingham and whose members are supposed to roar like lions when they rise to speak.

Certainly James Clerk Maxwell (1831–79), the first professor of the newly built Cavendish Laboratory at Cambridge, was a lion among mathematicians. His was the almost unbelievable triumph of proving that wireless must exist, more than twenty years before the first wireless signal was detected; of showing that this wireless must consist of waves; and of "measuring" the speed at which these waves must move—all, as I have said, before wireless as such was discovered.

It is the sort of success which a mathematician may achieve perhaps once in a century. Newton watched an apple falling, watched the moon's motion in the sky, and saw that both could be explained by a common force, gravitation. Einstein built up a more complicated and difficult theory and predicted, amongst other things, that the course of light would be bent when it passed near a star.

4

James Clerk Maxwell

Newton, Maxwell, Einstein—all three were true adventurers. They wanted to see things that no man had ever seen before. But whereas the new things that Einstein saw are far removed from the world as we know it, the abstruse calculations of Newton and Maxwell proved intensely practical. And alone of all the world's mathematicians Maxwell can claim to have founded a great industry.

Picture him in his study in London, still a young man as the world reckons to-day, keen, alert, with a reputation as much for wit as for learning, interested as mathematicians are in trying to find connections between things which to ordinary men seem unconnected. Always in a mathematician's work there is this idea of unification. It is his business to reduce everything to formulas, and, if only for his own convenience, it is obviously easiest to have as few formulas as possible. It is the mathematician who is the link, for example, between the "bite" of a skate on ice and the impossibility of boiling an egg on a mountain top. The "bite" of a skate depends on the momentary melting of the ice beneath the pressure of the skater's weight, although the temperature may be still several degrees below the ordinary freezing point; and whether an egg will, or will not, be cooked in boiling water depends on the temperature at which the water is boiling, which is in turn affected by the lower pressure of the rarefied air of the mountain top. So, dissimilar as these two effects may seem, they have at least this much in common. Each depends on a change of physical state, in one case that from solid to liquid, and in the other from liquid to vapour; and it so happens that both these sorts of change are affected by the surrounding pressure.

The connection which Maxwell sought was more ambitious. He wanted to weld all the known facts about light and electricity into a single coherent whole. It was to be the greatest unification which man had yet achieved, as great even as that represented by Newton's discovery of the law of gravitation.

Here again there was one definite similarity from which Maxwell could start. He knew that both light and the forces of electricity could reach out across space. We do not need telling that the headlamps of a car will light up the road ahead, or that it is the light from our distant sun which illumines both the earth and planets. And the very simplest experiment in electricity shows that electrical forces have the same sort of power. Most people must at some time or other have rubbed a stick of sealing-wax on their coat sleeve, so that it is electrically charged, and noticed that it will then pick up small pieces of paper just as a magnet will attract iron filings or a steel pin. In the same way the movement of an electric current in a wire will deflect a compass needle, and it will do this even if the needle is suspended in a vacuum. It is true that the scale is very different, but to this extent the effects are of the same kind. Both light and the forces of electricity can reach out across empty space, and Maxwell thought that they ought to reach out in the same sort of way.

It is difficult to-day to realise how miraculous the idea of "action at a distance" really is. It should be just as much a shock to a sunbather to find his skin

6

tanned by rays which have travelled more than ninety million miles through empty space as it would be to an aborigine to be knocked out by a white man's gun at a mere half mile. The aborigine's problem would be solved, if he was still interested, by the discovery that a bullet had been fired from a rifle; and that a bullet, presumably the same one, had lodged in his leg. He would no longer believe that the injury was magical, but would very properly conclude that the bullet had somehow passed from the rifle to where he was standing and was nothing more wonderful than a glorified edition of the arrows or slung stones which he himself used to kill game. But it is the bullet, and only the bullet, that removes the element of magic from the hurt which he has suffered.

So too, as science dislikes magic even more than does an aborigine, something had to be found to carry both light and the forces of electrical attraction through apparently empty space. The something was the ether, an intangible all-pervasive material which had been solely invented in the first instance to provide the vehicle for light's journeys about the universe. But exactly the same difficulty arose in the case of electrical forces, and Maxwell's essential idea was to make the same ether perform both jobs. To do this he had somehow to find a common factor in the light of the stars and the effect of a current-carrying wire on our compass needle. So he set to work with his equations to see what resemblances he could find. It may well have seemed a hopeless task. But in accomplishing it he proved that wireless, as yet undiscovered, was the necessary missing link between light and electricity.

The reason why there had to be wireless is not far to seek. Light was already known to consist of waves, so that if light and electricity were to appear at all alike there had to be electrical waves as well. The only difficulty is that most of us have no very definite ideas as to what is meant by any kind of waves except those of the sea. So before we can see what an enormous amount Maxwell was able to prove, merely on paper, about waves which no one had yet detected, we must make a short digression to find out what waves really are. As everything which wireless can do depends on its wave nature we shall not be wasting our time.

The simplest definition of a wave is that it is something which moves through space in a form which is regularly repeated. This is the characteristic, a little idealised, of the familiar invasion of sea rollers from the Atlantic. Occasionally we may see the ideal almost perfectly realised, each wave arriving at a regular interval after the last, each wave the same size, and each advancing at the same speed towards the beach. We may also notice that it is only the waves, and not the water itself, which are moving inshore. If the water were really, as it seems, moving towards us we should find that the tide on a rough day was always rising, and we ourselves would very soon be swept off the beach.

Without leaving our seat on the shore we can also, with the aid of a stop-watch and a little imagination, measure both the "wave-length" and the "frequency" of sea rollers—and can see how these two quantities are related. We might, for example, notice that the crests of successive waves were always about 400 feet

8

apart. That is what is meant by their wave-length. Then, bringing a stop-watch into action, we might observe that each wave was advancing towards the shore at the rate of 40 feet a second (2400 feet a minute), and that waves broke on the beach at a "frequency" of six a minute. It is no coincidence that the wave-length multiplied by the frequency of the waves is equal to their speed. It merely represents the rather obvious fact that if six waves, each 400 feet long, pass a given spot in the course of a minute, then the speed of the wave's travel is (6 times 400) feet per minute. This relationship is absolutely general and applies equally to any kind of waves.[1] It follows that the longer waves are, the lower will be their frequency, and *vice versa*.

Another general characteristic of wave motion is the capacity of one set of waves to interfere with another. If a stone is dropped into a pond we know that successive ripples spread out in a circle from the place where the stone hit the water. The "interference" effect is produced when two stones are dropped into the same pond at the same moment not very far apart. Each stone produces its own set of circular waves; and, although each set of waves will behave as if the other were not there, it will be impossible for them to avoid meeting and interacting. In some cases two troughs will coincide and produce an extra big trough; in others the trough of one set of waves will cancel out with the crest of the other; and in still other places there will be a combined crest.

[1] In this, as in what follows, it is irrelevant that water waves move only in two dimensions. They merely happen to provide the most convenient illustration of certain general aspects of wave motion.

How this works out in practice may be seen by drawing two circular sets of waves on a sheet of paper and marking the places where they meet, either to reinforce one another or to cancel out. The result, as shown in Fig. 1, is a set of nearly straight lines, which alternately correspond with the combined crests and troughs. Many beautiful photographs have been taken of the surface of a tank in which two sets of ripples, generally produced by the two prongs of a tuning fork, are seen "interfering" in this way.

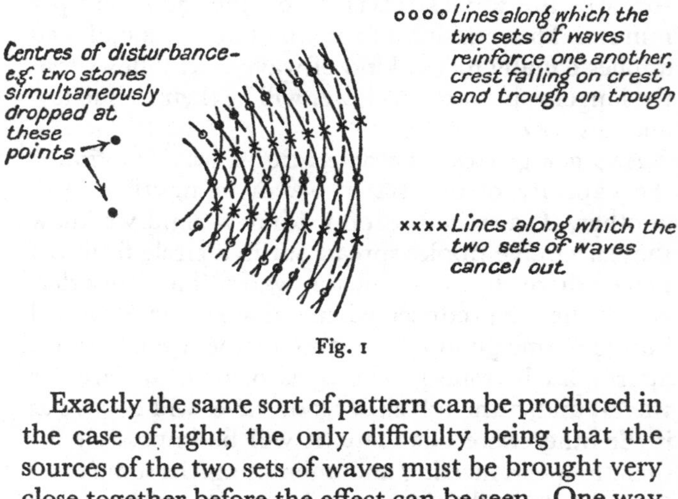

Centres of disturbance-e.g. two stones simultaneously dropped at these points

oooo Lines along which the two sets of waves reinforce one another, crest falling on crest and trough on trough

xxxx Lines along which the two sets of waves cancel out

Fig. 1

Exactly the same sort of pattern can be produced in the case of light, the only difficulty being that the sources of the two sets of waves must be brought very close together before the effect can be seen. One way in which this is done is by causing light from a narrow slit to fall on two mirrors joined together at a very fine angle. Two separate images of the slit are then formed, one in each mirror, and we can bring them as close together as we like by making the angle between the

mirrors sufficiently small. When the two images are
brought close enough we can see a series of light and
dark bands exactly corresponding with the lines of
greatest and least disturbance in the case of ripples.
The experiment is of rather special interest because
it provides one of the few ways in which we can
directly see light behaving as waves. The analogy with
similar patterns on the surface of water is perfect; and
it would not, in any case, be easy to imagine any other
way in which light, plus more light, could produce
darkness. It is possible to calculate the wave-length
of light from the distance between the light and dark
bands in these interference patterns. Red light has
the longest waves, and violet the shortest. But the
wave-length of any kind of light is almost infinitesimally small. That of yellow, which comes roughly in
the middle, is about one forty-thousandth of an inch.

So, watching first one kind of wave and then another, we can build up a picture of the generalised
character common to all members of the wave family.
From watching ocean rollers we have gathered that
all self-respecting waves are regular in their habits, as
well as the meaning of wave-length and frequency;
and by dropping stones into a pond that waves can
be mutually interfering. For their next trait we will
turn to sound waves which are of a different kind again
from any whose behaviour we have so far examined.

Sound waves tell us very conveniently how waves
are born. We have only to strike a tuning fork to see
that, so long as its note is sounding, the prongs of the
fork are in rapid vibration. In the same way it is
obvious enough, when a violinist plucks one of the

strings of his instrument, that there is a direct connection between the movement of the string and the
production of the sound; while the more varied tones
of an organ depend on the action of a series of metal
"reeds", which are free to move back and forth as
each pipe is sounded. In each case there is first
vibration and then sound; and, although the nature
of the vibrating source is not always so obvious, it is
none the less true that all waves originate in the same
sort of way.

How does this help us towards the possible production of radio waves? Very considerably, because any
ordinary alternating current is nothing more or less
than a centre of electrical vibrations. Such a current
is always waxing and waning in strength, never staying
the same from one moment to another. It first increases to a maximum in one direction, then reverses
and increases to a maximum in the opposite direction.
In the case of a standard mains supply there are fifty
cycles of change a second, successive changes in the
current being as regular as the purely mechanical
vibration of the string of a violin.

Maxwell's first great achievement was to prove
mathematically that any alternating current not only
might, but must, produce radio waves. Every electric
cable is all the time acting as a radio transmitter,
although the waves which it produces are so long that
we can never hope to recognise them as such. No one
can be expected to tell that a wave is a wave if he is
always inside it.[1] That is why Maxwell was the pro-

[1] It is a common misconception that there is some sharp dividing
line between alternating currents which do, and do not, throw out
waves. The above is the truer picture.

phet rather than the discoverer of wireless. Currents of very much higher frequency were needed to produce waves short enough to be detected, and in Maxwell's day the necessary experiments had not been made. But the correctness of every one of his conclusions is all the more impressive for that reason.

Thus did Maxwell predict the formation of wireless waves. But, as I have already said, he did more than that. He also calculated the speed at which they would move, and to his delight he found that this speed was identical with the speed of light. He was delighted because he had set out, you may remember, to prove that light and electricity were connected. It is also a particularly useful discovery from our point of view, for the simplest way to think out how we should expect wireless waves to behave is from what we already know of light. Wireless waves are only longer light waves than we are used to, so that if we look carefully at the scale differences between long and short water waves we shall have gone a long way towards obtaining a general picture of the "go" of wireless.

This part of Maxwell's work is as near an approach to a conjuring trick as can be expected from a mathematician. As with most conjuring tricks a little technical knowledge is required to appreciate to the full the skill of the magician's hand. But watching the extraction of rabbits from a hat is notoriously a pleasant pastime, and we can at least follow some of the main movements in what will always be regarded as Maxwell's star turn. There is the added fascination that Maxwell's conjuring trick begins and ends in reality, and that the methods used are at every stage above reproach.

He started from the known connection between electricity and magnetism, to which a passing reference has been already made. A compass needle, it will be remembered, is deflected by an electric current and will therefore change its position whenever the strength of the current alters. But the point which now concerns us is that the relationship is reciprocal. When a magnet is moved about, an electric current will begin to flow in any conducting material, such as a wire, in the neighbourhood. So that it is equally possible for a scientist to begin by studying electricity and to pursue his researches into magnetism; and to take magnetism as his starting-point and pursue his researches into electricity.

Electricity in Maxwell's day was therefore in the same sort of position that currency is to-day. Just as rival economists have joined battle over the commodity versus the gold dollar, so there were two rival systems of electrical measurement. One set of scientists, working on electricity, had drawn up one set of electrical units in terms of which they could measure the various quantities met with, such as charge, current and so on; and another set of scientists, working on magnetism, had drawn up a second set of electrical units to fit in with the magnetic units they had already chosen. It would have been as surprising if the two sets of units had agreed as if the yard and metre, having been independently defined, had been found to be of exactly the same length. What was more than surprising was

Heinrich Hertz

that the difference, unlike commodity-gold bickerings, proved of practical advantage.

Using his mathematical genius to probe the meaning of this discrepancy, Maxwell was able to prove that the ratio of one kind of electrical unit to the other (as it were the yard to the metre, or the pound to the kilogram) should be related to the speed at which the then undiscovered waves moved. Making the most accurate possible comparison of the two sets of units he concluded that the speed of the waves should be about 300,000 kilometres a second—as near as maybe the speed of light. This provided the best possible vindication of Maxwell's idea that light and electricity were connected, and also enables us to assume that light waves and radio waves are of the same kind. It is the most perfect example in all history of the power of a mathematician's pen.

Maxwell's calm assumption of the existence of wireless waves was verified, as the world now knows, by Heinrich Hertz (1857–94) in 1887, twenty-three years later. Maxwell was content to triumph in the realm of ideas. Hertz made the necessary tests with electric currents of higher frequency. He was able to show that the relatively short waves so produced could be reflected and could "interfere" with one another—in fact that they behaved as waves ought to behave. But transmission was only across a distance of a few feet in a laboratory, and no wireless valve was yet available to magnify the response at any distance into a reasonably strong signal. Nevertheless, the new form of communication had definitely arrived, and the new waves were duly christened after their discoverer, although it is hard not to feel that Maxwell's name would have been equally appropriate.

Although wireless and light were predicted and found to be of the same nature a common-sense proof that wireless consists of waves may be worth giving. Like most common-sense proofs it will not be water-tight. But the line of argument may be indicated like this. When an engineer at a transmitting station wants to send a message on a particular frequency what he does is to arrange that electrical vibrations of the frequency required take place in his aerial. Neither he nor the receiver knows anything directly of what happens to the wireless in the air. But the receiver does know that he has to "tune" his set until it is capable of giving electrical vibrations of the same frequency. Therefore receiver and transmitter, having compared notes, deduce that, whatever the wireless was in the air, its frequency remained unaltered during transmission. And something which moves through space in a form which is regularly repeated is what we have agreed to call a wave.

Naturally one of the most important parts of a wireless engineer's duties is to keep his transmitter working at the exact frequency allotted for the use of his station. There is an international laboratory at Brussels, the chief duty of which is to see that this is done. Regular records are made of the frequency at which different stations are broadcasting, and if any station starts breaking the rules its director very soon hears about it. Without this informal discipline there would be far more complaints from listeners of interference with one station by another.

One of the ways in which the frequency of a transmitter is kept constant is by making use of the fact that various crystals have natural periods of electrical vibration which they maintain with extraordinary accuracy so long as the temperature is kept constant. Quartz crystals are generally used, the frequency characteristic of the individual crystal depending on its size and the way in which it is cut. This cutting is a fine art, particularly in the case of crystals for use at short-wave stations, where the crystals have to be considerably smaller, as we should expect if they are to keep pace with the extra rapid vibrations which are necessary.

These crystals are the pocket ruler with which the wireless engineer measures the frequency of his station. But even the best of pocket rulers is not much use without some standard with which to compare it. The British standard of wireless frequency—which in turn is compared at intervals with the national standards of other countries—is an electrically maintained tuning fork which is kept at the National Physical Laboratory, Teddington. The electrically maintained part only means that electricity is used to keep it vibrating continuously over long periods.[1] To all intents and purposes it is an ordinary tuning fork. Its function is to maintain vibrations over a long period in an electrical circuit, but the mere fact that a tuning fork is used as the standard of both wireless and musical notes serves to emphasise very effectively the wave nature of wireless.

So far, however, this idea of waves has not taken us very much farther towards discovering how wireless gets round the world. A storm in the Pacific does not,

[1] It does this to within an accuracy of one part in ten million.

even after the longest interval, throw up waves on the coast of Cornwall. Nor is a searchlight, uplifted over London, likely to be visible in New York. At the same time it is obvious that there is a good deal of difference between these two kinds of wave, and it should not therefore be altogether surprising if their behaviour is not in all respects identical. Knowing that wireless waves are only light waves on a different scale, we will turn for a few pages to water waves, and see what differences we can find between waves of long and short wave-length.

We may, for example, make waves in a bath by dipping a nail-brush in and out—and these waves will be of the same kind as those of the sea, just as wireless waves are of the same kind as those of light. The difference is simply one of scale, and it is a very easy matter to show that scale effects are important. Thus a line of concrete posts widely spaced along a beach, is enough to break up the large-scale movement of sea rollers and so save the shingle from being washed away; and if the posts were sufficiently big, they would not only help to break up the waves but would act as an efficient reflector, sending them back where they had come from, out to sea.

Unfortunately there would not be room for so big a post in our bath, and the best imitation which we can give is to use one of our own legs instead. Then, if we stoop down and make some ripples with the nail-brush, we shall find that our leg is powerless to stop their even flow past it on either side, although immediately behind there would be a calm area which we might recognise as the shadow of our leg. If we wanted to reflect the ripples we should have to use a "mirror" of very much finer grain. A long comb, held in the water, would do the job nicely. Here, then, are two particulars in which we should expect the scale effect to be significant—the sort of obstacles past which the waves can move unaffected, and the sort of mirror which is necessary to reflect them.

Finally, there is a third scale effect for which it is easiest to go to a harbour mouth. There on a fine day we may see a slow procession of ripples moving in past the pier heads—even the most complete landsman could hardly call them waves in any nautical sense of the word. All the harbour will be shielded from their advance by the piers except the part immediately in front of the harbour mouth, and if we watch we shall see that there is very little spreading of the waves to one side or the other. The section which can come in through the entrance moves inwards very nearly in a straight line.

Now choose another day when there are some real waves about, and see how they enter the harbour. This time every boat will be rocking at its moorings. No part of the anchorage escapes the disturbance of the advancing waves. They are smaller than they were to start with, but instead of moving forward in a straight line from the mouth, they spread out more nearly in the form of a circle. In fact, whereas the smaller waves travelled nearly in a straight line the bigger waves can bend round a corner. Not, of course, that any kind of wave cannot bend round corners to some extent. We could hardly imagine them as waves if they did not. Given a toy harbour and a toy-sized

19   2-2

harbour mouth the ripples would have behaved in precisely the same way. But they want a different kind of obstacle to make them bend round corners. This is the third scale effect which is important.

Great as is the difference between the lengths of ripples and long sea rollers, the difference between the wave-lengths of light and broadcasting is even greater. The waves of yellow light, it will be remembered, are about one forty-thousandth of an inch long. In comparison the waves of radio wear seven-league boots. The broadcasting range, as the dials of our receiving sets tell us, is in round figures from 200 to 2000 metres—or if we prefer English units, from 220 to 2200 yards. The longest broadcasting waves are therefore about three thousand million times as long as those of yellow light, so much longer that the imagination boggles at the comparison. So it will not be surprising if we find the scale differences between ripples and sea rollers considerably magnified in the case of light and wireless. In addition the range covered by radio waves is very much greater than that covered by light waves, so it will also not be surprising if radio waves differ more among themselves. The table printed below shows a comparison between the lengths of long, medium and short waves in radio and light respectively.

Wireless wave-lengths in metres		Light wave-lengths in units a thousand million times smaller	
Long	3000–20,000	Red (longest visible)	720
Medium[1]	100–3000	Yellow (medium)	580
Short[1]	10–100	Violet (shortest visible)	400

[1] In the classification adopted by the International Radiotechnical Committee an additional category of "intermediate" waves is intro-

Having obtained the scale relation between light and wireless we are now ready to guess at the sort of differences in their behaviour which may be expected. In the case of water waves, it will be remembered, scale differences showed themselves in the extent to which waves moved in straight lines, the sort of obstacles past which they could move unaffected, and the sort of objects by which they were reflected. If these differences are reproduced in wireless—as they are—it will help to make the behaviour of wireless more credible.

Ripples, we saw, moved more or less in straight lines, while long sea rollers had a greater capacity for bending round corners. In exactly the same way light, being of relatively short wave-length, moves for all practical purposes in straight lines, and casts shadows. But wireless is very much better at negotiating any obstacles which it finds placed in its way. We know that while a hill or a mountain range does not improve wireless reception, it seldom makes reception absolutely impossible as would be the case with a light ray.

The longer wireless waves are, the more different they are from light and the more we should expect them to bend round corners. Equally, the shorter wireless waves are, the more likely will they be to travel more or less in straight lines. In practice it has proved that the limits of straight-line travel fall within the range of wave-lengths now available for wireless—

duced. All wave-lengths between 200 and 3000 metres are regarded as "medium"; between 50 and 200 as "intermediate"; and between 10 and 50 as "short". While there is much to be said, on technical grounds, for the addition of the intermediate category, the meanings to be attached to the terms "medium" and "short" waves in the following pages are those given above.

a state of affairs which is not, on the whole, to be regretted since it makes possible a very much wider range of radio technique. Thus the long and medium wave bands used in ordinary broadcasting can bend round obstacles fairly completely, while short waves (which means anything below 100 metres) are sufficiently directional to be aimed at particular receiving stations or parts of the world.[1] The fact that they travel more nearly in straight lines means that energy need not be wasted in broadcasting these waves in all directions when reception is only wanted in a particular region. This is the idea behind Marchese Marconi's beam wireless and the similar stations established by the British Post Office at Rugby and Baldock. Since their object is to provide a regular wireless telephone or telegraph service between England and selected receiving stations abroad, transmission can be made as strictly directional as is humanly possible.

The B.B.C. Empire services are also directional, the longest wave-length used being 49 metres to Canada. But here there is the difference that one service is expected to serve not one station, but anything up to a continent. The problem of getting adequate, and preferably fairly even, reception over a large but none the less limited area is quite a new one, and, as the B.B.C. engineers are the first to admit, has not yet been completely solved.

[1] There is also the purely practical factor that with the smaller aerials of short-wave transmitters it is possible to provide a sufficiently extended aerial system to concentrate the waves into a beam; as also to double the intensity of the beam in the forwards direction by employing a second or "reflecting" array behind the first. See also Chapter III.

So-called ultra-short or micro waves—which mean anything from 10 metres downwards—move, as we should expect, still more nearly in straight lines, and are therefore still more economical for communication between fixed points. I shall have more to say about the future of these waves in a later chapter. In the meantime it may be said that they also share, to a greater extent than do other types of wireless, in another property of light—that of finding brick and mortar a serious obstacle. They will penetrate a brick wall, but find a block of high buildings an effective barrier. More will be known about this when the B.B.C. has completed its television experiments on a 7·75 metres wave-length from Broadcasting House, but experience in other countries has already shown that the difficulty is a real one. On the other hand the Post Office's wireless telephone service across the Bristol Channel on a similar wave-length has given complete satisfaction, and by the time this appears in print the original transmitter will have been replaced by new equipment designed to carry six conversations at the same time on wave-lengths only slightly separated. In this and other ultra-short wave telephone services the two stations are naturally chosen so that there are no obstacles in between.

This is one example of the second scale difference between light and wireless—their capacity for negotiating obstacles, without which indeed wireless would be of enormously reduced usefulness. The same effect is also met with in the way the energy of wireless is absorbed in travelling along the ground. We normally expect that a beam of light shall either arrive at its destination or be completely blocked. But when the headlights of a car shine along a road it is obvious that a large part of the light is scattered, and at the same time some of the light is absorbed as energy by the road.

In the same way wireless loses a lot of energy in a long journey during which it often passes near or hits the ground. The distance to which a "short" wave can reach travelling directly over land is only about 50 miles, although the loss of energy depends a good deal on the type of soil. For some years now the B.B.C. has conducted regular soil surveys round new stations in order that the loss from this cause shall be as small as possible. Many of these samples have been examined at the National Physical Laboratory at Teddington, and the same institution has made direct measurements of the travel of wireless of various wave-lengths over different types of soil. A damp soil is better than a dry one, and clay than gravel. To its soil Droitwich owes, at least in part, its fame as the successor of Daventry for home broadcasts.

The range of a direct wireless wave being as small as it is the scope of wireless would have been seriously curtailed if nature had not kindly provided a way out. But since the discovery of how wireless gets round the world is inevitably linked with the practical achievements of Marchese Marconi, Marchese Marconi must come first.

The Marchese Marconi

CHAPTER III

RADIO ROUND THE WORLD

THE story of wireless is the story of man's annihilation of distance, and in its practical development no man has played a greater part than Marchese Marconi. Maxwell prophesied, Hertz discovered, Marconi gave the increase.

As a young man of twenty, the idea became firmly rooted in Marconi's mind that in these new-fangled waves lay the possibility of communicating over great distances. He started his experiments at his father's house at Pontecchio, near Bologna, in 1895,[1] and succeeded in receiving signals at a distance of a mile. It does not sound anything to get excited about to-day—but remember that the original triumph of Heinrich Hertz was measured in feet.

Marconi's success was largely due to his realisation of one fact. He saw that to produce strong signals he must provide a long length of aerial in which his electrical vibrations could take place. It is a train of thought which leads directly to the earthed aerial, the main practical development with which Marconi is generally credited. It might, of course, have been theoretically possible to get the required length of

---

[1] Any complete history of radio would include at this stage, the names of Sir Oliver Lodge, and of Preece of the British Post Office. It is extraordinary how nearly Lodge, in particular, missed the honour of being the pioneer of radio communication. But to Marconi, in any case, belongs the credit of practical achievement.

aerial by fixing it horizontally, say, between two trees, in the same sort of way that receiving aerials are generally arranged. But Marconi soon found that, when an aerial of that kind was used for transmitting, the waves which it produced were only too readily absorbed by the ground. The aerial must therefore be vertical; and if the aerial is to be vertical there are other reasons why it should also be "earthed".

What Marconi was doing, it must be remembered, was making a source of rapid electrical vibrations communicate their movement to a wire, so that as much as possible of the energy produced should be radiated out into space. He was in the position of a man bowing a violin to make the string vibrate, or of a boy (which he very nearly was) jumping up and down on a plank bridge to make the bridge move as violently as possible. He could make the bridge move appreciably at whatever spot he elected to jump, but he would certainly produce the best results if he jumped somewhere near the middle of the bridge. So it was desirable that Marconi's electrical machine, which was to be the original source of his vibrations, should be placed in the middle of the aerial. That would have been easy enough if the aerial could have been slung horizontally, but in a vertical aerial it implied a crow's nest type of construction which even the most enthusiastic of experimenters would naturally want to avoid if possible.

The solution which he evolved is both simple and effective, even if the problem did not appear so clear-cut at the time. He left his electrical machine at ground level, where he could look after it without difficulty, and merely led the bottom end of his aerial into the earth. He found that the wire then behaved like a double-length aerial, of which the second and imaginary half is buried in the ground. His signals were therefore as strong as he could make them, and it is for the same reason that every outdoor receiving aerial is to-day connected to "earth"—or to a metal pipe leading down to the earth, which comes to the same thing.

Marconi was so enthusiastic that the next year he came over to England, took out the first wireless patent, and succeeded in interesting the British Post Office in his ideas. So it came about that some of the earliest practical experiments on radio were made from the roof of the General Post Office in St Martin's-le-Grand. After some further demonstrations to the Post Office on Salisbury Plain, Marconi betook himself, also with Post Office support, to the Bristol Channel. Here, between a hill at Penarth and Brean Down near Weston-super-Mare the range of wireless was successfully raised to ten miles. It is a historic bit of coast this, from the point of view of long-distance communication. Here also the Admiralty made their first trials of signalling by heliograph, and here to-day is the first ultra-short wave telephone circuit in this country. It crosses practically the same bit of the Bristol Channel as did Marconi's signals.

Two years later Marconi conquered the English Channel and radio received its first practical applications. Within the same year radio was first used in naval manœuvres; by Queen Victoria to send messages during Cowes regatta from Osborne House in the Isle of Wight to Prince Edward on the royal yacht; and during the Kingstown regatta in Ireland to provide newspaper reports for the Dublin *Daily Express*. Then in October 1900, a very much more powerful station was built at Poldhu in Cornwall, and at the very first attempt Marconi succeeded in receiving signals on the far side of the Atlantic at St John's, Newfoundland. It was development intoxicating in its success—and in one sense intoxication was the result. Marconi's trans-atlantic waves were relatively long compared with the laboratory transmissions of Heinrich Hertz. The very ease and swiftness with which long-wave communication progressed was largely responsible for nearly 30 years' virtual neglect of short waves.

In this his greatest triumph, Marconi got all the credit of doing the "impossible". At the very time that the transmitter was crackling away at Poldhu and a jubilant Marconi was watching his new receiver at work in Newfoundland, mathematicians were busy proving that wireless reception over distances of more than a few hundred miles was impossible. It was not possible, they said, for wireless waves to scatter sufficiently to get round the curvature of the earth for any greater distance.

To-day we are in a happier position. We have no need of abstruse mathematics to prove that wireless waves make no straightforward progress round the earth. Every listener can learn from his own experience that this is not the case. The most striking demonstration is provided by the B.B.C.'s Empire station, and the most homely by fading. Everyone with any interest in

28

Empire communications will remember the excitement which followed the first of these transmissions from Daventry on a series of short wave-lengths. Home listeners with short-wave sets immediately wanted to listen-in to the Empire programmes, and in spite of warnings by the B.B.C. they attempted in many cases to do so. What happened was that no one more than 50 miles from Daventry was able to hear anything at all. On the other hand, listeners more than 500 miles away—for example in the extreme north of Scotland— began once again to get reception. In between was a completely "dead" area, with not a sound to be heard except the intermittent crackling of tuning-in.

Looking at these results it is difficult to avoid the conclusion that the waves received 500 miles from Daventry had, in the meantime, travelled sufficiently far up into the atmosphere to be well out of the reach of any receiving set—in other words, that they had not pursued a direct course round the surface of the earth.

Fading in reception, which must have been experienced by every listener, leads to the same conclusion. With conditions on the earth's surface apparently much the same, and no change in either transmitter or receiver, why should reception suddenly fade to nothingness, and then after a brief interval be restored to normal? It is hard not to feel that somewhere between transmitter and receiver the fugitive wave must have visited some region where conditions are more liable to sudden change than is the surface of the earth.

At the time every effort was naturally made to explain Marconi's success without introducing any new terms into the scientific coinage of the day. It is easy enough to invent a genie to carry wireless round the world, or to do anything else that seems difficult or impossible. But one of the chief aims of science is to keep its genie population as small as possible, especially when—as in this case—there appeared to be no very immediate hope of being able either to prove or disprove the existence of the genie.

The chief of these efforts depended on the known bending of light as it passed from air into water or glass—that is, from a less dense into a more dense medium. It is this bending which gives an appearance of eccentric twist to an oar, seen half in and half out of the water. It is also the basis of the power of a lens to concentrate light into a focussed beam. In fact, it is something quite different from mere scattering, the possible limits of which had already been explored.

In the case of the atmosphere, the change in density was very simply provided by the smaller pressure at great heights. Briefly, the idea was that the top part of a wireless wave would travel a shade more quickly in the rarer atmosphere above. It would therefore be gaining all the time on the lower part of the wave. So the combined front of the wave's advance would be tilted more and more forwards, and in this way the wave would be bent round the earth. The only question was whether the bending could be big enough to serve the purpose.

It was at this point that this delightfully simple theory broke down. Sir Ambrose Fleming, one of the veterans of British radio science, calculated that only if the earth's atmosphere consisted entirely of the rare gas krypton, would the bending be of the right amount. And the results, as he pointed out, would be, to say the least, disconcerting. Apart from the minor drawback that we should be quite unable either to breathe or live in such an atmosphere, we could, if we had a powerful enough telescope, look half-way round the world and watch the progress of a test-match in Australia; or even, in moments of extreme exuberance, look the whole way round and inspect our own backs. In fact, this otherwise attractive theory would explain a great deal too much. Light would be bent round the earth's surface, just as wireless would be. Sailors would never have known the limits of a horizon, and it would be possible to send even the shortest wave wireless round the world—which engineers know only too well that it is not.

The true explanation—that there were one or more "mirrors" in the sky which reflected wireless round the world—was put forward by Oliver Heaviside (1850–1925), the fourth of the great personalities of wireless and the man from whom the "Kennelly-Heaviside" layer takes one-half of its name. He was an English telegraph engineer, who to the great benefit of science was compelled by increasing deafness to retire from professional life at the early age of twenty-four. An eccentric, living a completely secluded life at Torquay, he worked out not only the most vital part of modern wireless theory, but the whole theory of long-distance telephony as well. A. E. Kennelly, who independently suggested the existence of a wireless "mirror" in the sky, was an American professor of English descent who is still alive.

What sort of mirror had these two men in mind? What sort of mirror could there be in the sky which could conceivably reflect wireless waves? Heinrich Hertz, it is true, had proved that his much shorter wireless waves could be reflected. But this was done with a piece of wire gauze—which was hardly helpful towards the solution of Marconi's problem.

If there was such a mirror, would it keep still enough to be of commercial service? Could it always be relied upon? These questions, answered at first by the dead-weight of practical experience, raise some of the most fascinating questions in the whole story of wireless. Perhaps it is as well that Marconi and his helpers did not wait to answer them before getting on with the job. The fact of wireless had to come first, although the "how" of wireless directly affects every single one of the world's 150 million listeners.

Once again, we have to go back to light, to the sort of mirror that we all instinctively think of. It is the old story of the "scale difference" between light and wireless. Wireless waves, which are unfamiliar, are somewhere about a thousand million times longer than light waves, which we know well. How should we expect the longer waves of wireless to be reflected? We noticed that different kinds of objects would reflect ripples and sea rollers. But if we are to obtain the clue we need to the behaviour of wireless, a rather more detailed comparison is necessary. And the first thing we want to know, as our "mirror" has got to be in the sky, is how solid it must be.

From the analogy of light we are used to thinking of a mirror as something completely solid. But even in the case of light we may remember that the atoms which make up the reflecting surface are full of holes. And in the case of water waves it is obvious that there can be large gaps in the surface of a perfectly efficient mirror. We have seen, for example, that a comb will reflect the small ripples we make with our finger in a bath; or a piece of wire gauze, in which the holes are much more extensive than the wire, will serve the same purpose. And, if we go again to the seaside, we may see the same sort of effect on a very much larger scale.

Standing at the end of a pier, built up on a network of girders, we may watch a small rowing boat tossing and pitching as it rounds the pier-head. The waves, just off the pier, are no longer regular, but confused and tumbled as if they did not know which way to move. The same movement can be felt, no less alarmingly for passengers in a small boat, if the boatman ventures in too close beneath a tall cliff which sheers suddenly into deep water. In each case the waves are being reflected, at pier or cliff face; and where the reflected waves meet the incoming rollers behind is a characteristically choppy stretch of water.

The point is that both the solid cliff face and the pier's network of steel girders produce the same effect. Yet ripples, the wave-length of which is small compared with the spaces between the girders, would pass on unperturbed beneath the pier. That is one essential condition for the reflection of any kind of waves. Any gaps in the "mirror" must be small compared with the length between successive waves. The other essential condition is that any bumps in the mirror must be small judged by the same standard. A light mirror must be polished to an extraordinary degree of smoothness, but a sheet of wood is smooth enough for the sounding board above a pulpit. A cliff face will reflect rollers, or throw back a sound wave to form an echo, but its jagged boulders are far too rough to reflect the waves of light.

Wireless waves being so much longer than those of light, we are therefore at once entitled to expect that they will require something much less solid to reflect them. We may see this at the B.B.C.'s Empire station at Daventry, where the reflector used to prevent the waves spreading backwards as well as forwards is nothing more substantial than a second aerial system, placed some twenty feet behind the first. Even more striking, because on a smaller scale, Marchese Marconi uses in his ultra-short wave transmitters what may be roughly described as a semicircle of metal rods.[1]

Even so, we do not expect to find either metal rods or aerials floating about in the sky, and we must look more closely at what happens in light reflection if we are to see what kind of sky mirror we may expect to find. The atoms which turn back light waves are no more solid than is the solar system. On the same analogy they have a relatively dense nucleus, which corresponds with the sun, and a number of electrons moving round it which we may call the planets of the atomic world. And since the electrons, which are only particles of electricity, are spread over a very much

---

[1] More accurately a parabola, the necessary shape to produce a parallel beam.

larger area than the nucleus, it is presumably the electrons which first make their influence felt on any invader of the atomic world. But the way in which the planetary electrons treat an invading light wave is hardly the behaviour normally expected of a defending army. The light wave is given the very politest of rebuffs.

We may imagine that each electron in the path of the waves is set dancing to the same tune as the light wave which is attempting to pay them a call. In other words, they all start vibrating at the same frequency as the light wave. Each electron in its dance holds part of the energy of the light wave for an infinitesimal fraction of a second. Then it hands its quota of energy back again in the form of a new light wavelet. Finally, when all the electrons have completed their fleeting dance of welcome, all the separate wavelets which each has formed combine together to produce the reflected wave. The electrons, by their sympathetic, apparently welcoming dance, have sent the light wave out again whence it came.

With surfaces of other kinds than that of a mirror, the welcome may be more real. If a searchlight, for example, is played on the surface of water, part of the light will be reflected exactly as if the water were a mirror. But another part of the light is bent sharply downwards at the surface and passes on to illuminate the bed of the stream. This means that the electrons at the surface of the water have communicated their dancing urge to those farther down, and so the light is carried down beneath the surface. The two processes are very much the same. Each depends on a momentary dance of countless electrons, and when we come to speak in greater detail of wireless mirrors in the sky, it does not matter very much if we think of the light as being reflected or bent. Even an ordinary light mirror really bends the wave gently back on its course. It is only because the scale of a light-wave is so infinitesimally small that we think that a light ray which hits a mirror has been sharply reflected at a point.

In our comparison of light and wireless with ripples and sea rollers, the reflection of light corresponds with that of ripples; and the network of electrons which turns light back on its course with the wire gauze which we saw was sufficiently solid to reflect ripples. Just as sea rollers could be reflected by the much more open structure of a pier-head, so we shall expect to find a wireless wave less exacting in its requirements. But because wireless and light are of the same kind, we shall still look for a network of electrons to serve as our mirror, even though the network be millions of times less closely knit.

As it happened, scientists were already aware of a type of looser electron network which exactly fitted the bill. Only, in fact, when there is such an electron network will a gas conduct electricity. But, until Heaviside and Kennelly suggested it, no one had suspected that it had any connection with wireless.

I said before that the planetary system, or outer part, of all atoms consists of electrons. Electrons are also the unit particles of electricity, and when we say that an electric current is "passing" we mean that a stream of electrons is being passed from atom to atom just as fire-buckets are passed from hand to hand along a chain of helpers. That, at any rate, is what happens in a wire or solid conductor, where the atoms are close on top of one another and can easily play their part in the chain. But the particles of a gas are much more widely separated, and cannot under ordinary conditions take part in any co-operative movement of this kind.

In order to pass the message of electricity from one atom to another the particles of a gas are forced to send out runners—and the runners they send out are their own planetary electrons. That is to say, a certain proportion of the gas particles are split, or "ionised". The result is that the gas not only contains its own heavier particles, but free, unattached electrons. These electrons are free not only to move across the air gap in an electrical circuit and so to carry the current; but to dance to any kind of tune which a wireless wave passing their way may ask of them. That is the connection between the passage of an electric current through a gas and the reflection of wireless. In each case the electrons must be free to move about on their own.

There are various ways in which a gas can be "ionised", or made electrically conducting. One of these is by the passage of an electric spark through it; another is by heat, as with the flame of a candle; and still another is by the sort of rays that radium gives off. We may say that all of these things "knock spots off" the atom—but that as the spots are neither very big nor very firmly attached, the knock does not have to be very violent. For the same reason quite a number of different kinds of "knock" produce the same effect.

Here, then, is a possible kind of electron network which might act as a wireless mirror. We are then faced with the question, is it likely to be close enough knit for the purpose? Why should it be formed in the sky? And is there any other reason to suppose that it exists?

As to whether the network is close enough knit, we can obtain a rough idea without actually plunging into arithmetic. Wireless waves, it will be remembered, are somewhere about a thousand million times longer than light waves. We expect, therefore, to be able to reflect wireless with a network which is a thousand million times less close than that of the planetary electrons in the solid surface of a mirror. The first step on the downward path is to notice that the atoms in a solid are much more closely packed than those in a gas—a contrast which is simply conveyed by the fact that the one is so much heavier than the other. Steam, for example, is about 1700 times lighter than water.

Again, when a gas is "ionised", each particle of the gas can add only one electron to the network, whereas the atoms of a solid mirror can each muster a whole army of planetary electrons to assist in the repulsion of the light wave. Next we must remember that the air high up in the atmosphere, where the "mirror" presumably is, must be even less dense than at ground level; and, finally, that only a small proportion of such gas particles as still remain can be expected to be "ionised" or split up in this way. There are four stages in this thinning-out process—we make our atoms more scarce by proceeding from a solid to a gas; we only allow each atom to contribute one unit to the network

38

instead of many; we make the network even coarser by taking it high up into the sky; and even then we have to admit that only quite a small proportion of the atoms are playing their part. The extent to which the network is coarsened in the first two stages we could, if we wished, work out exactly. For the remaining two stages we should have to indulge in a certain amount of guesswork. But the guesswork need not be altogether wild and, as a matter of fact, we should have no great difficulty in bringing the answer out to the thousand-million-fold increase in coarseness which is what we are looking for.

The sort of "mirror" we have been talking about is therefore capable of doing the job. The answer to the second question—why should it be formed in the sky?—is also satisfactory. Gases, we noticed, are not very particular about the kind of knock they are given when someone wants to "ionise" them. At the same time we know that the sun's radiation must be very much more intense high up in the atmosphere than it is at ground level, because it has not had to come through any of the thicker layers down below. So there is no difficulty in imagining that the sun's radiation may be strong enough to give the required knock. Later we shall be able to see exactly how it does it.

Finally, the answer to our third question—is there any other reason to suppose that such an electron network exists?—is "Yes, there is, and there was reason to suspect it the whole time". More than ten years earlier it had been suggested that an electrically conducting layer high up in the earth's atmosphere was needed to explain small recurrent changes in the

39

magnetism of the earth. And it had also been pointed out that such a layer would go some way towards explaining the wonderful displays of aurora borealis noticed by all Arctic explorers. The aurora is fairly obviously, like lightning, electrical in origin; but, unlike lightning, only shows its light high up in the sky. So it is not difficult to see that scientists might find the existence of an electrically charged layer of air at the same sort of height, useful in explaining it. At the same time, neither the magnetic eccentricities of the earth, nor the aurora, could be regarded as urgently pressing problems, and so the suggested electrically charged layer of air had remained—a casual suggestion.

The whole position was altered by the further suggestion of Heaviside and Kennelly that such a layer was needed to explain the world-wide progress of wireless. Wireless was not something which could be conveniently ignored, and in any case the ideas of Heaviside and Kennelly were backed up by a formidable array of mathematics. But no direct proof was supplied. And as scientists are never willing to accept anything new without proof, a sharp dispute raged during the whole of the 22 years from 1902 up to 1924 —which is the next important date in the story of how radio gets round the world.

Shortly before, Professor E. V. Appleton, at that time a research worker at the Cavendish Laboratory, Cambridge, had seriously turned his attention to the wireless wave. He brought to the task the particular combination of qualities for which that laboratory is famous—first-rate mathematical equipment and a capacity for practical experiment which can only be described as brilliant. While his earlier work at Cambridge does not concern us, Professor Appleton was attracted in 1924 to the idea of providing definite and direct proof that such a wireless "mirror" existed. How he set about it takes us back to the analogy of water waves. It will be remembered that when two sets of ripples met they "interfered", waves from each set of ripples either reinforcing one another or cancelling out, according as crest coincided with crest, or crest with trough; and that precisely the same effect could be produced with light.

Professor Appleton's idea was to apply the same sort of method to wireless. Assuming that the Kennelly-Heaviside layer existed, he expected, from a single transmitter, to receive two separate sets of waves at any receiving station, one set having pursued a direct course along the ground, and the other having been up to the Heaviside layer and down again. Obviously the one which had been up and down again would have travelled farther than the other, although since wireless travels as fast as light he could not hope to make any direct measurement of the time interval between the two sets of waves. He was driven, therefore, to make use of an "interference" effect of the type indicated. As we shall meet the same, or very similar methods being used in later investigations of wireless "mirrors", we may as well look at it in some detail.

It will be simplest to take a practical example. Let us suppose that wireless waves are travelling from transmitter to receiver by two routes, one 300 and the other 200 miles long, the difference between the lengths of the routes being just 100 miles. Confining our attention to the shorter or direct route, we must imagine that the whole space between transmitter and receiver is filled with an unending succession of advancing waves—crest and trough, crest and trough, and so on until the receiver is reached. Now suppose that the wave-length of our transmitter is so adjusted that the distance between receiver and transmitter is occupied by an exact number of wave-lengths. A wave-length, it will be remembered, is the distance between crest and successive crest, or between trough and successive trough; so if, at any instant, the transmitter is sending out a crest there will now, at the same instant, be another crest arriving at the receiver.

Now let us have a look at what is happening on the indirect, or 300 miles, route to the receiver. In the first place, we know that 200 miles from the transmitter a crest will at the same instant be arriving, because we have chosen our wave-length so that this will be the case. But whether a crest will be also arriving at the receiver will depend on whether the extra hundred miles distance, on the longer route, can also be divided up into an exact number of wave-lengths. If it can, there will be a crest at the receiver and the direct and indirect waves will reinforce one another; but if there is a half wave-length over, then the indirect wave will produce a trough at the receiver which will cancel out with the direct wave's crest.

How can we tell when this will happen? It depends on whether an even or an odd number of wave-lengths was needed to fill exactly the 200 miles' distance of the direct path. If an even number of waves was needed, then when we divide this number by two to fill the 100 miles excess distance of the longer route, the answer will still be a whole number of wave-lengths. And so a crest will be arriving at the receiver by each route. But if an odd number of wave-lengths was needed to fill the direct path, we shall have a half wave-length over when we divide it by two, and so the two waves will cancel out at the receiver.

Finally, let us put in some actual figures for wave-lengths and see how it works out. Our direct distance is 200 miles, so let us start by making each wave exactly a mile long. Two hundred waves will then just fill up the direct route, and three hundred the indirect route, and the two sets of waves will reinforce one another at the receiver. Now let us make the waves just a little bit longer so that 199 waves will just fill the direct route. That means that each wave is 1·005 miles long. Without doing any complicated arithmetic we can see that one and a half times 199 waves, that is, 298½ waves, will be needed to fill the longer route, and that at the receiver the two waves will cancel out. We could have produced the same effect by making our wave-length 0·995 of a mile, so that 201 waves were required to fill the direct route. In that case, 301½ waves would be needed for the indirect route and, again, the waves would cancel out at the receiver. Equally, going a step farther on either side, we should find that wave-lengths of 0·990 and 1·010 of a mile once again resulted in mutual reinforcement by the direct and indirect waves.

So, running rapidly through the whole series, we

have—wave-length 1·010 miles, waves reinforce; wave-length 1·005, waves cancel; 1·000, reinforce; 0·995, cancel; 0·990 reinforce. Thus if the wave-length was being continually changed we should expect to get alternate periods of good and bad reception. Equally there is a direct connection between the change in wave-length which is necessary to produce this result and the difference in length between the long and short routes. With the figures chosen a wave-length difference of 0·005 of a mile is the change needed to produce an aggregate difference of half a wave-length when the waves are spread out so as to fill a hundred miles.

Such was the method used by Professor Appleton to demonstrate the real existence of the Kennelly-Heaviside "mirror" and to measure its height above the earth's surface. To carry it out he betook himself and his receiving equipment to Oxford in the summer of 1924, and enlisted the services of the B.B.C.

The broad outline of his experiment is almost childishly simple. The B.B.C. engineers at the Bournemouth transmitting station undertook to vary the station's wave-length after the close of the programme, through a range of 10 metres on either side of the then normal wave-length of 386·5 metres. Professor Appleton, listening in the electrical laboratories at Oxford, had nothing more to do than to watch out for any unevenness in the strength of reception. He found, as he had hoped and predicted, that as the wave-length changed he got alternate periods of strong and weak reception. From the change in wave-length between one peak of good reception and the next, he could calculate the difference between the lengths of the two routes taken by the radio waves which were "interfering"—and the schoolboy's old friend, Pythagoras, did the rest. Fig. 2 shows that given the distance between Bournemouth and Oxford, and the difference between the direct and reflected routes, it is only a matter of the geometry of a right-angled triangle to work out the height of the "mirror".

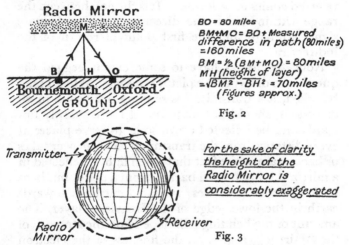

BO = 80 miles

BM+MO = BO + Measured difference in path (80 miles) = 160 miles

BM = ½ (BM+MO) = 80 miles

MH (height of layer) = √(BM² − BH²) = 70 miles (Figures approx.)

Fig. 2

for the sake of clarity the height of the Radio Mirror is considerably exaggerated

Fig. 3

After 22 years the existence of the Heaviside layer had been proved and its height above the earth charted. The mere fact of the regular variation in reception as the wave-length was changed proved its existence, and the rate at which the variation took place provided a measure of its height. We can now for the first time draw a picture of a wireless wave going round the world (Fig. 3).

It will be noticed that the picture shows a wave leaving the transmitter at such an angle that it hits the mirror fairly obliquely. The effect of this is that the indirect ray, in its bouncing course round the world, does not make its first return to the earth's surface until it has travelled quite an appreciable distance. The picture, therefore, gives an indication of how the "dead area" of radio reception to which I have referred comes to be formed. It is the gap between the range within which the direct ray gives adequate signal strength and the first point of return of the indirect ray.

None the less, I have to some extent begged the question in drawing the picture this way. Why, it may be asked, if a transmitter is sending out waves in all directions, should not some travel more sharply upwards and be reflected down so as to serve places at every distance from the transmitter. The answer takes us back again to the fact that when light is reflected in a mirror it is not turned back as from a point, but bent back. Just so, a wireless wave is bent back towards earth in the lower edge of the Heaviside layer. The amount of the bending will depend on the strength of the "mirror", that is on the fineness of the electron network. It will also depend on the wave-length of the radio, because it will be remembered that the longer the wave-length the less dense the network which we should expect to be necessary to reflect it.

These ideas explain practically all the known facts about "dead areas" of reception. It is obvious that the more oblique the reflection from the mirror the less violently will the wave have to be bent back in order to return again to earth. A wave travelling vertically upwards needs to be bent back through a full 180 degrees, while a very much smaller amount of bending will bring back a wave which hits the mirror obliquely (Fig. 4).

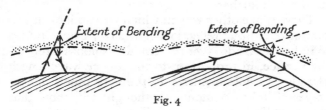

Fig. 4

This, then, is the explanation of dead areas. We expect that, unless the "mirror" is strong enough to reflect all waves of a particular wave-length, irrespective of the angle at which they hit it, a certain degree of obliqueness will be necessary before the wave is at all strongly reflected. And so, after all, our original picture of a wireless wave going round the world was justified. None the less, there is rather more to it than that. Very careful records made in America have shown that the extent of the dead area depends very markedly on wave-length. The shorter the wave-length the longer the dead area, as these figures show:

Wave-length	Length of dead area (mid-day, summer)
16 metres	1000 miles
21 ,,	600 ,,
32 ,,	300 ,,
40 ,,	200 ,,

This, again, fits in perfectly with theory. The longer wave-lengths not only require a less powerful mirror

47

to reflect them, but can be bent back more sharply by a mirror of the same strength. The indirect wave can therefore make an earlier return to earth on a longer wave-length, and the dead area will be still further shortened owing to the fact that the direct wave will itself reach farther.

There is one other way in which we can look at this matter of oblique reflection. Hitherto, I have talked as if the Heaviside layer presented as smooth a reflecting surface, in proportion to the longer wave-length of radio, as does a mirror to the very much shorter waves of light. A moment's thought will show that this is not likely to be the case. The air, even at ground level, is never still. Even more, higher up there must be eddies and currents which will prevent our electrically charged layer from keeping either a smooth or sharply defined surface. In fact, it will be much more like a sheet of paper than a mirror in the strict sense of the word; for even the smoothest of paper shows corrugations under a powerful microscope, and is very far from flat from the point of view of anything so small as a light wave.

How does this make good reflection more probable at an oblique angle? Have you ever tried holding a sheet of paper up to a light, so that paper and light are almost in line? The paper which has before looked dull, begins to shine brightly, and the more nearly paper and light are brought into line the more nearly will the paper behave as a mirror. If eye and paper are sufficiently nearly in a straight line, it is even possible to see a definite image, as of the filament of an electric lamp.

48

The same effect is to be seen on a wet road at night when the headlights of a car, shining almost straight along the road, are reflected as in glass. Just in the same way we should expect the presumably rough surface of the Heaviside layer to be a more efficient reflector of rays that hit it obliquely. So we must think of the earth's globe as being surrounded by an outer skin which behaves like a cross between a mirror and a sheet of paper. For the longer wave-lengths of wireless, it behaves rather more like a mirror. It will reflect them at whatever angle they approach it, but as from any rough surface there will always be a certain amount of scattering. But for shorter wave-lengths the outer skin behaves more like a sheet of paper. It will only reflect waves which hit it at quite a fine angle.

Finally, there is one respect in which the wireless-reflecting skin of the earth behaves neither like a mirror nor a sheet of paper. Indeed there is nothing in the reflection of light which is quite comparable. The mirror can be *too* strong. By this I do not mean that there can be too many electrons in the electron network, but rather that the "free" electrons may not have sufficient elbow room. The electrons have momentarily, it will be remembered, to take up the dance of the wireless wave. If they are always banging into other air particles the energy given to them by the wave is absorbed before they have had time to hand it out again as the reflected wave. In other words, the wave will be absorbed instead of reflected. It will pass into the Heaviside layer and be no more heard. This is quite an important possibility, the effects of which on radio reception we shall see later.

CHAPTER IV

UP IN THE SKY

PROFESSOR APPLETON'S success in proving that the Heaviside layer really existed was spectacular and important. But it left many questions unanswered. He had shown that this wireless mirror was in action at a particular place and time, and that it would reflect waves of a particular wave-length. It remained to be shown whether it always stayed at the same height, whether its strength varied or was always the same, and what were the limits of its capacities.

It was necessary that the newly found wireless mirror of the skies should be kept under continuous and varied observation if the many problems were to be solved, affecting engineer and listener alike, to which it apparently held the key. Many watchers, an organised programme and a special transmitter were required if the job was to be done thoroughly. In England the last two requirements were met by the Radio Research Board, a government organisation formed in 1920 under the Department of Scientific and Industrial Research. The object of the Board was to provide for research of just such wide scope, and a special transmitter, which could be varied over a wide range of wave-lengths, was erected at the National Physical Laboratory at Teddington.

Of watchers there was first and foremost, Professor Appleton, now established at King's College, London, and himself a member of the Board. A second receiving station was available at Peterborough, where the Post Office's research station had already been loaned to the Board, and the co-operation of Professor Appleton's former colleagues at Cambridge was also secured. It was thus possible for experimental transmissions to be made from Teddington, and to be simultaneously received in London, Cambridge and Peterborough.

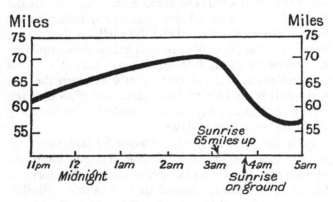

Fig. 5. Record of one night's changes in the height above the earth of the Heaviside "mirror".

The first step, obviously, was to keep the Heaviside layer under sufficiently prolonged observation to show whether any day and night changes took place in its height. During the early stages of the investigation, working on a wave-length of about 400 metres, records were limited by interference from broadcasting stations. Fig. 5 shows the "log" of a typical summer night's record.

The figure shows the height of the "mirror" gradually rising, reaching its highest point shortly before 3 a.m., and then falling sharply to smooth out at a daylight level of about 60 miles above the earth. Such early records have two great points of interest. They prove in the first place that the variation of height in the mirror is by itself enough to be of practical importance. They also provide a very strong indication of what it is that keeps the radio mirror active. In the figure it will be noted that one arrow indicates the time of sunrise 65 miles above the earth on the day the record was taken, while a second arrow shows the time of sunrise at ground level. As the period of most marked change falls almost exactly between the two arrows, it is difficult to avoid the conclusion that the return of the sun's rays is responsible for the lowering of the radio mirror at dawn.[1]

Although there are other reasons for believing that the sun is connected with wireless, it is possible to see at once how sunrise and sunset come into the story. It has already been explained that the sun's radiation may be responsible for "knocking spots" off the particles of the upper atmosphere, so as to form the network of free electrons which is the Heaviside mirror. Therefore it is only natural that, when the sun's bombardment is removed, some of the wandering electrons should meet some of the particles from which they had been so rudely separated and agree to reunite. The unromantic light of day brings a spate of broken marriages. The return of night brings the erring

[1] A marked difference between day and night reception had been noticed by Marconi as early as 1902, the same year that the existence of the Heaviside mirror was suggested.

electron back, if not to her original husband, at least to someone else's.

The nearer the free electrons are to the earth, the more likely will they be to run into a particle with which to join forces, because the denser the air will be. So, as the barrage of the sun's radiation is progressively lifted, it is only to be expected that the lower part of the network of free electrons should be gradually depleted by such random matings. Higher up the process of reunion will be slower, because there will be fewer opportunities for liaisons, until somewhere about 70 miles above the earth's surface, as the figure shows, a nearly steady population of free electrons is left as the lower edge of the Heaviside mirror.

The day and night watch on the radio roof of the world was destined, however, to provide very much more startling and more important results than these. Although the figure already reproduced was typical of the records obtained on many nights, there were other occasions, particularly in winter, when it seemed to the watchers that the apparent height of the wireless mirror was suddenly doubled in the hours shortly before dawn, dropping again to its normal level when day came. Suspicion at first fell on the accuracy of the records. A sudden discontinuity of this kind appeared most improper and unscientific, and the most careful measurements were undertaken to determine whether the effect was a real one.

One Sunday night's records at King's College clinched the whole matter. Professor Appleton had to work on Sundays because it was the only day in the week when his laboratory was free from the usual

double bombardment of disturbing noise from the
Strand and the Embankment. But for this particular
Sunday's work he had good reason to be grateful.
Fig. 6 shows the record of the apparent height of the
wireless mirror shortly before and shortly after dawn.

How was this apparently reliable "log" to be explained? Clearly it was fantastic to suppose that the
"mirror" had been suddenly lifted some 70 miles,
after being comparatively steady before, and then as

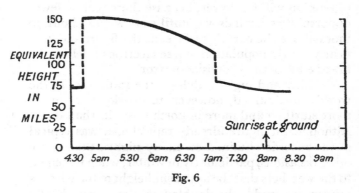

Fig. 6

suddenly dropped back again. A second possible explanation was that at the moment of the jump the
receiving station was, for some reason unknown, suddenly ceasing to get the normal reflected wave, which
had been down and up once, and was getting instead
a twice reflected wave which had been up and down,
and then up and down again. Some rather similar
records obtained in America were, in fact, interpreted
in this way. But what Professor Appleton found could
not be so lightly dismissed. The objection to this simple

54

explanation was that the apparent height of the mirror
should in that case be exactly doubled. Fig. 7 illustrates
the path of a wave which has been twice reflected
between Bournemouth and Oxford, and shows that
this path will have exactly the same length as that of a
wave which has been once reflected twice as high.

Professor Appleton's measurements, it will be remembered, showed that the apparent height of the
mirror suddenly jumped up from 75 to 150 miles. So
far, then, as the upwards jump was concerned the

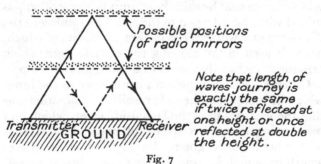

Fig. 7

explanation that the waves in question had been twice
reflected at the now familiar Heaviside mirror might
perhaps have been accepted; although, as Professor
Appleton pointed out at the time, there was no very
obvious reason why once-reflected waves should not
in that case have been received as well. But the downwards jump, later in the morning, from 110 to 75 miles
was in quite the wrong proportion, and the extent of
both upwards and downwards jumps varied percept-
ibly on different days. The great merit of his work was
therefore that he was prepared to vouch for its accuracy,

55

and on the strength of his belief to commit himself to the existence of a second radio mirror, normally at rather more than twice the height of the first.

His obstinate and wholly justified belief in his own work has an interesting parallel in the story of chemical discovery. More than 30 years earlier the great scientist, Lord Rayleigh, had found that nitrogen obtained from the air was heavier by about five parts in a thousand than nitrogen prepared in the laboratory. Like Professor Appleton he was so certain that his work was accurate that he said there must be some other gas mixed with the nitrogen of the air. The result was the discovery of a whole series of rare gases, one of which, neon, is used in the glowing red electric signs which ornament Piccadilly Circus and tobacconists' shops.

Given Professor Appleton's confidence in the jump which he measured, the only possible explanation was that the weakening of the Heaviside layer, which had been going on progressively all night, had at that particular moment got to the stage when the wavelength on which he was working could just pierce it. Instead of being reflected by the first wireless mirror, the waves then travelled farther up through the atmosphere until they met a second reflecting layer higher up.

There is another and even more spectacular way of watching the jump from one mirror to another. It depends on being able to do the very thing which I said that Professor Appleton could not expect to do— that is, to time the odd thousandth of a second difference between the arrival of the wave which has travelled straight along the ground and of the wave which has been up to one or other of the mirrors and come down again.

To do this a transmitter is used which at intervals of one-fiftieth of a second sends out almost infinitesimally short "jabs" of energy. Each "jab" lasts one-ten-thousandth of a second. Yet so fast does wireless travel that when the tail end of the last wave of one "jab" is just leaving the transmitter, the front of the first wave has already advanced nearly 20 miles on its journey. And in the odd ten-thousandth of a second the transmitter, working in this case on a wave-length of 100 metres, has sent out no less than 300 successive waves—quite a big enough sample to test which of the two wireless mirrors is in working order.

The separation of the direct and reflected waves at the receiving station involves the use of one of the most remarkable instruments of modern science, the cathode-ray oscillograph. As this instrument is undoubtedly destined to play an important rôle in many of the future applications of radio, including television, it is worth a little attention. In spite of its formidable name it is nothing more than a convenient way of watching a stream of electrons being pulled hither and thither by whatever forces we care to expose them to. The electrons are wanted in this case, not because of their capacity for forming a wave-reflecting network, but because they are the lightest sort of particles known to man. The slightest push will send them flying to one side, just as a beginner on a bicycle has an infinite capacity for wobble. But there is this difference. The "cathode-ray oscillograph" recovers as easily and as quickly from a wobble as it gets into one.

For the benefit of watcher or camera the beam of electrons is shown as a green spot of light on a fluorescent screen. It is generally arranged that the arrival of these small convoys of wireless waves makes the spot jump upwards; and if our eyes could follow jumps made fifty times a second with any comfort that would be all that was needed. There is, however, a very simple way out of this difficulty. All that is necessary is to arrange that the spot travels quite independently from left to right, jumping suddenly back again, and that it executes this movement in tune with the transmitter, that is fifty times a second. Any upwards jump of the green spot, which is regularly repeated at this interval, will then always appear in the same place.

We can now see how the cathode-ray tube helps the scientists in their watch on the radio mirrors. Since the transmitter is sending out its brief batches of waves at intervals of a fiftieth of a second, successive batches of waves, whether they have travelled along the ground or have been reflected, will arrive at the receiver at similar intervals. But there will be a small interval, a thousandth of a second or a little more, between the arrival of waves from the direct and indirect routes. So the arrival of the direct waves will be shown as one fixed peak in the line traced by the green spot, and the arrival of the reflected waves as a second peak a little distance to one side. Also, if the second set of waves have been up to the Appleton instead of only the Heaviside layer, they will be farther behind the first set and the two peaks will be more widely separated.

What happens when the "jump" from one mirror to the other is taking place? The watcher can see the

58

first mirror gradually weakening, as if it was tired of reflecting wireless waves, while the upper mirror makes its first appearance on the screen, takes on a larger and larger share of the work, and is finally left in sole charge. The change-over can be seen in every stage, just as a film of a relay race shows the first runner coming up strongly, then the old and new runners moving side by side, and lastly the new runner carrying on by himself. The baton has been passed.

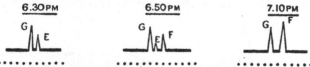

Each dot represents ⅟₁₀₀₀ᵗʰ second interval
G. Arrival of ground or direct wave
E. Arrival of wave reflected at Heaviside mirror.
F. Arrival of wave reflected at Appleton mirror.

Fig. 8. The change-over from one radio mirror to the other.

Fig. 8 shows three stages in a radio change-over. They were taken at intervals of 20 minutes, the transmitter being at East London College and the receiver at King's College, Strand. The dots underneath represent time intervals of a thousandth of a second. *G* stands for the arrival of the ground wave, *E* for the wave reflected at the Heaviside mirror, and *F* for the wave from the Appleton mirror.

It will be seen that in stage (1) the reflected wave is all Heaviside, while in stage (2) the Appleton mirror is beginning to take fairly complete charge and by stage (3) has completely cleared the field. In the original pattern, as a matter of fact, a very small notch

59

could even then be seen between the G and F peaks, corresponding to the last gasp of the Heaviside mirror. But to all intents and purposes the Heaviside mirror had been completely supplanted within the space of this 40 minutes.

The next picture (Fig. 9) has also a very special interest, since it is obviously necessary that a radio wave should be reflected more than once between earth and sky mirror if it is to get very far round the world. This picture shows the arrival of as many as

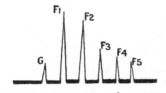

$F1,2,3,4,5$ represent the arrival of waves which have been reflected once, twice etc. up to five times from the Appleton mirror.

Each dot represents $\frac{1}{1000}$ second interval

Fig. 9

five successive reflected waves, reflection having taken place in each case at the Appleton mirror.

Looking at the dotted "time scale" underneath, we can see that the Appleton mirror must in this case have been at a height of about 180 miles. That means that each reflection must have corresponded with a journey of some 360 miles. So in order to get 5 miles from East London to the Strand the last wave of the series had travelled no less than 1800 miles. The five upward and downward paths must have been so close to one another that it would be quite impossible to draw a clear scale diagram showing how this wave travelled. But we have only to imagine such a diagram telescoped out sideways, in the same way that an extensible shaving mirror can be pulled out, to get a clear picture of how wireless gets round the world.

There is one other point of interest about this picture. It will be remembered that American workers thought that the "jump" to the Appleton mirror corresponded only with two reflections from the mirror which was already known. It is now clear that there was nothing fantastic or improbable about this explanation—except that Professor Appleton's measurements were accurate enough to show that it did not in all cases fit the facts.

The tools were now at hand. It was up to the scientists to make use of them to chart the positions and strengths of these two mirrors at all times of the day and night. Obviously the way in which they moved and their changing capacity for reflection was of the very greatest importance for every one concerned with wireless.

The stage of adventure was over, and since routine records tend to be dull, I will merely summarise what the scientists found from their watch. In the first place the reflecting edge of both the mirrors tended to move up round about sunset and down again about sunrise. This, it will be remembered, is because when the sun's bombardment is lifted the electrons composing the reflecting network begin to attach themselves again to the particles from which the sun's radiation has knocked them free. As there are more particles lower down where the air is denser, it is the lower edge of the layers that melts away most rapidly and completely.

Broadly speaking, the Heaviside layer is about 10 miles higher by night than by day, although there may be odd days when the layer is fairly steady throughout the twenty-four hours. But the Appleton layer is very much more lively. In fact, except around mid-day, it is hardly ever still. It has been recorded as low as 90 miles in Australia, has a usual day level of about

Miles

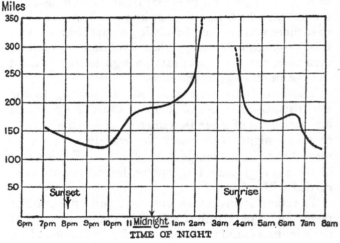

Fig. 10

120 miles, and at night may soar up to well over 300 miles as Fig. 10 shows. It will be noticed that there is a gap in the small hours of the morning, indicating that the greatest height of this layer has never been explored.

We shall see later that the mirror whose erratic movements are depicted above is the one on which all short-wave broadcasts are chiefly dependent. When we remember that engineers are trying, by repeated reflections from this most wobbly mirror, to steer a transmission from London to Australia, the wonder is not that service is erratic but that reception is possible at all.

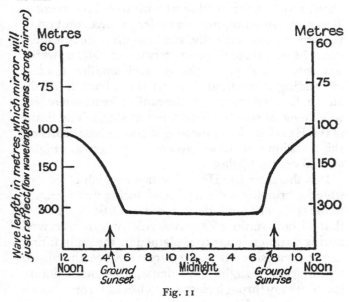

Fig. 11

Naturally it is even more important to know whether the mirrors will be in working order for the wave-lengths used at any particular moment. This will depend on the strength of the mirrors, as measured by the shortest wave-length which each is just capable of reflecting. Fig. 11 shows the shortest wave-lengths

63

which the Heaviside layer will just reflect at different times during a winter's day. The arrows show the times of sunrise and sunset at ground level.

It will be noticed that during the hours of darkness the Heaviside layer will not reflect radio of shorter wave-length than 300 metres, and that even at noon about 110 metres is the lowest with which it can cope. These are winter figures[1] and refer to waves sent up to the layer nearly vertically. On long-distance transmissions the waves are not despatched vertically upwards but only at a slight angle. A much smaller amount of bending is required to bring them back to earth, and under these conditions the radio mirrors can reflect wave-lengths roughly four times as small. The limits of the Heaviside layer for long-distance broadcasts on this occasion were therefore about 75 metres at night, and 28 metres by day.

It is therefore the Heaviside layer which is responsible for bringing ordinary broadcasting wave-lengths to the receiver's set under normal conditions. The fact that distant stations can as a rule only be received during darkness is partly explained by the slight lifting of the Heaviside mirror at sunset. Since the air 10 miles higher is slightly less dense, absorption due to loss of energy through electron collisions is correspondingly reduced. What is probably more important is that, at the same time as the mirror rises, its rather "fuzzy" lower edge is cleared away. This is an "absorbing region", a downwards extension of the Heaviside mirror which is only strong enough to absorb radio without reflecting it. During the night, when

[1] But not necessarily applicable to *any* winter day.

this barrier is no longer there, wireless waves come immediately to a part of the layer which is strong enough to reflect them. Less of their energy is therefore frittered away in advance, and so the paradox arises that although the Heaviside mirror is weakest at night

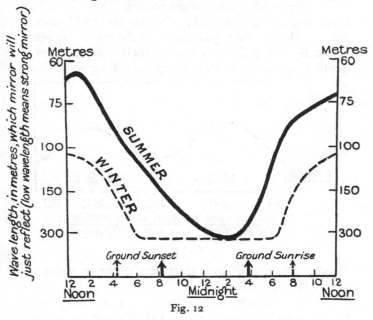

Fig. 12

it is most effective at that time. In other cases some of the waves may pierce the Heaviside mirror, and be reflected at the Appleton mirror, where there is still less absorption.

Fig. 12 shows the contrast between summer and winter conditions. The dotted curve, showing how the critical wave-length changes during a winter's day, is reproduced exactly from the last figure. The continuous curve shows an exactly similar record for a summer's day.

It will be seen that at its lowest point the summer curve just touches the winter record. There is still a time, shortly after midnight, when the layer will reflect nothing shorter than 300 metres, if the waves are sent vertically upwards, or than 75 metres at the more gentle slope of waves that are going a long way. But

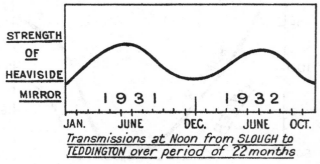

Fig. 13

at noon the mirror can reflect long-distance waves of about 18 metres.

Finally, this part of the story may be completed by yet another figure (Fig. 13) showing how the "strength" of the Heaviside mirror at noon changes with the seasons. It is very nearly a two years' record, the years covered being 1931 and 1932. No such complete record is available for the Appleton layer at the time of writing, although it has been proved that it is in all respects very much more lively.

66

Such records show that the short-wave services which now span the world are largely dependent on the Appleton mirror. Nature has kindly provided the Appleton mirror as a second line of defence to bring recalcitrant wireless waves back to earth. In fact, if only the wireless mirrors would stay still, the arrangements provided would be almost ideal. Comparison of the wave-lengths which each of these two mirrors will just reflect shows that the Appleton layer is just about four times as strong a reflector. Waves which are too short for the Heaviside layer can go on travelling upwards and make use of the Appleton layer. On the other hand waves with which the Heaviside layer can deal adequately have no need to make any farther journey.

The Appleton layer having been waiting patiently in reserve the whole time, it may at first sight seem surprising that short waves as a practical means of long-distance communication were such relatively late arrivals to the wireless world. The fact is that Marconi's startling success with long waves cast everything else into the shade. Hertz and indeed Marconi himself, in his early experiments, had generated short waves, but they had been practically forgotten by all save scientists.

In the end it was the amateur, driven to experiment on rejected wave-lengths, who did as much as anyone to secure the rehabilitation of short waves. American stations using wave-lengths of less than 200 metres began to be heard in this country in December, 1921, and soon after amateurs in different parts of the Empire began to indulge in the pastime of distant conversation on considerably shorter wave-lengths. Some of these short-wave stations, though operating with very small power indeed, were heard across 7000 miles of the globe with only a simple two-valve receiver. During the winter of 1922–3 Dr W. H. Eccles and Mr H. Morris-Airey, both English amateurs, were able to listen to the Pittsburg broadcasting station on 91 and 80 metres, as well as to a number of American amateurs. But it was left to an English schoolboy to span the greatest distance of all. In October, 1924, Mr C. W. Goyder, then at Mill Hill School, established communication with New Zealand, the farthest possible point on the earth's surface.

To these enthusiasts must be given part at least of the credit for the short-wave renaissance. It was they who made plain the almost universal usefulness of short waves for long-distance communication, even if Marchese Marconi had also an important finger in the pie. Marconi, who had investigated the military potentialities of short waves for the Italian Government during the War, made an experimental voyage to the West Indies in 1923, which convinced him that short waves should be capable of providing a commercial service to South America. Trial short-wave transmissions had been begun a year before from the Poldhu station, Cornwall, from which the Atlantic had first been spanned 20 years earlier, and it was these transmissions which Marconi was able to pick up in his yacht *Elettra*. The same experiments proved that, with a suitable aerial array, it was practicable to concentrate waves of up to about 100 metres into a "beam" which could be aimed in any desired direction. Although many different arrangements have been used, the result resembles the focussing of light by a mirror. We have only to think of how much illumination would be needed to give the whole sky the brightness of a searchlight beam to realise the saving of electrical power effected by the beam wireless system. It is to-day the basis of practically all commercial communication on short waves.

Finally, there is one other chapter in the story of wireless reflection, which deserves a mention, though it is of no practical importance. It has been stated earlier—and this is generally true—that all wavelengths which are too short to be reflected by the Appleton layer pass out into space and are no more heard. Just occasionally, however, stray wireless waves return to the earth after an interval of, perhaps, minutes—and a minute, for a wireless wave, means a journey of more than ten million miles. How they do it is one of the unsolved mysteries of radio; but the facts, which are beyond dispute, and decidedly interesting, are as follows.

The existence of wireless echoes, by which are meant weaker signals arriving appreciably after the main signal, has for a long time been recognised. But the most usual echo is a mere seventh of a second later, which exactly corresponds, at the rate of wireless travel, with a single circuit of the earth. No one therefore took any special interest in wireless echoes until a Norwegian engineer, Mr Jorgen Hals, reported in September, 1927, that he had heard a second echo a full 3 seconds after the first. If this vagabond wireless wave had spent the interval going round the earth, it would have had time for more than twenty circuits. Even if that was possible, there would be no obvious reason why there should not have been a separate echo heard for each circuit. But Mr Hals had only heard two echoes—if that explanation was correct, the first and the twenty-first. Therefore, if what he heard was real, a new theory of some kind was required.

The original 1927 echo was received while listening to signals from the omni-directional Dutch station at Hilversum, working on 31·4 metres. When similar echoes ranging up to 30 seconds were simultaneously recorded by a number of different observers at Oslo and in Holland there could be no doubt that something real had been discovered. Then a Norwegian scientist, Professor Størmer, to whom Jorgen Hals had communicated his results, was able to save the face of the puzzled scientists by proving that an existing theory of his own would fit the facts.

Professor Størmer had devoted infinite pains to calculating mathematically the observed forms of aurora borealis displays. These polar lights, he had concluded, were due to the arrival of streams of electrically charged particles from the sun. But since the earth behaves as a large magnet these particles would be deflected from a straight path as they came within range of the earth's magnetic forces. It was on this sort of basis that Professor Størmer had worked out how all the wonderful forms of the polar lights could be produced, and in the process he had decided that there ought to be irregularly shaped bands of electrons stretching far out beyond the earth's ordinary atmosphere.

*Aurora Borealis—one of many fine photographs
taken by Professor Størmer of Oslo*

Here, then, was a ready-made explanation. Some of these electron bands, almost out in free space but not quite, were dense enough to reflect short-wave wireless. But, alas for scientific complacency, the length of the wireless echoes recorded was speedily pushed up to include intervals too large for any reasonable explanation. The longest is apparently an echo of 4 minutes 20 seconds, reported by Mr Hals in 1929. Assuming that the wireless waves in question were travelling the whole time at their usual speed they would have had time for a double journey of more than twenty million miles.

The next step was to suggest that these obstinately eccentric waves might have been so far slowed up, while travelling in the Appleton layer, that they might have spent the necessary time on quite a respectably short journey. This explanation has, however, been proved to be untenable by Dr P. O. Pedersen, a Danish authority on the mathematics of wireless waves. He finds that wireless waves cannot be held up in this way for long enough to account for more than a 10 seconds delay in their arrival. Any longer interval would mean that the wave had been so far absorbed that there was nothing left to arrive. Nor can echoes of more than 30 or, at the most, 60 seconds be accounted for by reflection at Professor Størmer's electron bands.

The present position is therefore both intriguing and unsatisfactory. The waves which produce all the longer echoes must have been travelling at something approaching their proper speed. But scientists must in that case call in two different explanations of their

71

existence. Professor Størmer's earth-streamers account very prettily for the shorter long-delay echoes, and it is definitely in his favour that his theory was in existence before the facts were produced. But it is difficult to see how the longest echoes can be explained except by supposing that far away, out in space, there are, for reasons quite unknown, still other bands of electrons which are strong enough to reflect wireless waves. Even that is only putting forward a theory in order to ask a question, and the difficulties are so great that many scientists are inclined to doubt the genuineness of the longer echoes. How and why should these distant mirrors be so arranged that they not only reflect wireless waves back to the earth, but tactfully concentrate them into a beam similar to that produced by a directional station? On no other basis could the echo signal be strong enough to be heard. Yet, how and why? Science has no answer, nor looks at the moment like providing one.

Note: for the Scientifically Minded

For the sake of completeness it may be added that the structure of both the Heaviside and Appleton layers, which have been discussed earlier in this chapter, appears to be complex. Above the main Heaviside layer, generally denoted E^1, there is evidence of a secondary region of electrification, known as E^2. The region E^2 is generally "weaker" than E^1 and therefore plays no part in normal radio communication, since all waves which it is capable of reflecting will already have been reflected at the layer E^1 below. But in the early morning, when solar radiation is successively strengthening the reflecting powers of lower and lower regions of the atmosphere, there may be times when E^2 is stronger than E^1. A limited range of wave-lengths, capable of penetrating E^1 will then be reflected by E^2.

In the same way there is evidence of the existence of a "protuberance" or "ledge", described as F^1, on the lower side of the main Appleton layer, F^2. Since F^1 is more strongly ionised than the Heaviside layer it may reflect radio waves under normal conditions. It is, however, less strongly ionised than F^2. Its existence on many occasions during daylight hours is therefore indicated in radio records by the fact that the apparent height of the Appleton layer may be greater for longer than for shorter wave-lengths reflected at it. In other words the longer wave-lengths which can penetrate the Heaviside layer appear to be reflected at a somewhat lower level than the main Appleton layer. Fig. 14 illustrates diagrammatically the height and strength of successive radio-reflecting layers in the atmosphere.

There may also be two higher reflecting layers, which have been provisionally christened the *G* and *H* regions. Reflections from both these mirrors are relatively weak. The lower boundary of the *G* region is described as being commonly about 380 miles above the earth, but the height of the *H* region may be as great as 1100 miles.

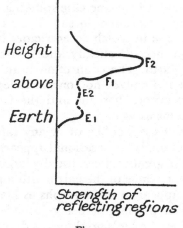

Fig. 14

73

CHAPTER V

THE SUN CALLS THE TUNE

DISTRESSING as are the day-and-night and winter-and-summer changes of the radio mirrors from the point of view of wireless engineers, the sun cannot be held responsible. It is the motion of our own planet that is to blame, in the one case spinning on its axis, in the other moving round the sun. There is, however, one genuine case in which eccentricities in the sun's behaviour do most drastically affect wireless conditions on the earth. The connection is an interesting one because it involves not only wireless, but the weather, the aurora borealis and the frequency of magnetic storms as well.

The cause of the trouble, or at any rate the most obvious symptom, is the occasional appearance on the sun's surface of great vortices, usually large enough to swallow up the whole of the earth with a good bit to spare. They appear in photographs as black spots of various shapes, the largest being just visible to the naked eye when the sun's brightness is dimmed by fog, or in the early morning or evening by horizon haze. They are by no means a modern discovery, the earliest Chinese records of their appearance going back to A.D. 188. Between that year and 1638 sunspots, as they are now called, make sporadic appearances in the Chinese Annals, being generally described as "fleckle in the sun" although sometimes the terms

74

"bird-shaped", "flying bird-shaped" and "like an apple" are used.

It was not until 1610 that sunspots were first noticed by the Western world, and their discovery then followed the invention of the telescope. It was by watching sunspots that astronomers first learnt that the great mass of the sun turns round on its axis about every 27 days as viewed from the earth; but although sunspots were frequently, if sporadically, observed during the next two centuries, no records were made during the whole of that time which were sufficiently regular to bring order out of the chaos of individual observations.

It was not until 1826 that the German astronomer, Schwabe, began a regular watch on the sun. Then, by sheer persistence, he discovered what all previous observers had missed—that the number of sunspots waxed and waned in a regularly recurring cycle of roughly eleven years' length. Finally in 1874 those early Chinese jottings of "fleckle in the sun" were at long last translated into a continuous modern record, kept at Greenwich Observatory, and soon after enlarged to include reports from other observatories in India and South Africa. The measurements obtained show that the sun has for more than a century behaved with commendable regularity. It is true that the interval between successive periods of greatest quiet has sometimes been as short as ten years and sometimes as long as thirteen. But on the whole the sun has stuck remarkably closely to the 11·4 years' interval which has proved to be the average length of its cycle of change.

A "period" of rather more than 11 years also fits in with the admittedly scanty early Chinese records, as well as with the still spasmodic observations contributed by seventeenth and eighteenth century astronomers in Europe. It is true that there was a period of 70 years from 1645 until 1715, during which the sun was either relatively inactive or the records unusually incomplete. But even then the same 11-year rhythm can be traced, and so far as we can judge the evidence suggests that the sun has been behaving in very much the same way for more than 2000 years. Fig. 15 shows part of the Greenwich record. Sunspots, in this case, are represented, not by mere numbers, which fail to distinguish between large and small spots, but by the average area of the sun's surface, which from day to day during each year was covered by spots. The observed areas are also corrected for perspective, according to the position of each spot on the sun, so that the figure probably provides as good a record of this form of solar activity as can be obtained. It will be seen that the cycle is very well marked, and that 1933 appears as the probable minimum of solar activity for the present cycle. The next maximum should be about 1938–9.

Meteorologists have devoted a great deal of time and trouble to attempts to trace connections between solar activity and weather on the earth. It is not so easy a task as it sounds because weather is so notoriously fickle. As Sir Napier Shaw has put it, there are so many syncopations in weather music, that it is difficult to spot the regular rhythms beneath. Most of all is this true of our own country. After investigating a

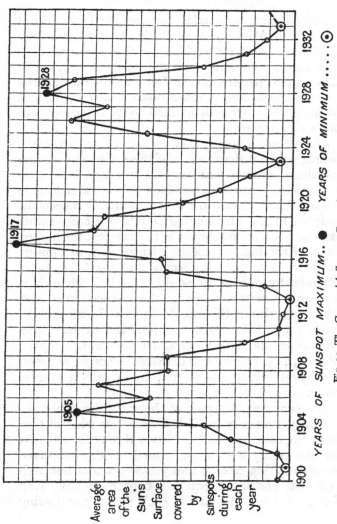

Average area of the Sun's Surface covered by sunspots during each year

YEARS OF SUNSPOT MAXIMUM .. ●   YEARS OF MINIMUM ..... ⊙

Fig. 15. The Greenwich Sunspot Record 1900–1933

hundred years of London weather Mr David Brunt of the Air Ministry, now Professor of Meteorology in London University, failed to find any significant rhythm to correspond with the sunspot cycle; and if any regular pattern does in fact underlie the vagaries of our erratic climate, it is probably safe to say that it is not sufficiently well marked to be of any use for forecast purposes.

There are, however, a few cases in other parts of the world where it does look as if there are definite connections between weather and sunspots. The most convincing example is provided by the water-level in Lake Victoria Nyanza, which presumably reflects the rainfall in the surrounding country. Here weather and sunspots seem to move in perfect step, the highest level of the lake coinciding with sunspot maximum, and *vice versa*. It also seems that there are more thunderstorms over the world as a whole when there are most sunspots, and that this connection is most clearly marked in Siberia, the tropical Pacific and the West Indies. The percentage increases in these cases are quite marked—Siberia, 16 per cent.; tropical Pacific, 43 per cent.; and the West Indies, 24 per cent. It is, however, in Siberia that the two rhythms seem to move in closest step.

To complete this part of the story the American scientist, Professor A. E. Douglass, has tried to turn back the pages of weather history through 3000 years by examining the rate of growth of the long-lived sequoias ("big trees") and yellow pines of northern Arizona. The oldest sequoia is believed to have lived for 3200 years, the size of each year's growth-ring giving a rough indication, so Professor Douglass argues, of the then rainfall in that particularly arid climate. There may be some doubt as to whether the growth-rate of the sequoias has been correctly interpreted, but the yellow pines alone are sufficient to summarise the weather of Arizona for the last 500 years—200 years earlier than the first Western observations of sunspots. Professor Douglass finds that their rate of growth also corresponds with an 11·4-year cycle, and is convinced that this is the sunspot cycle. As a further artistic touch Professor Douglass has also stated that the difference between "maximum" and "minimum" growth was very much less in the 60 years from 1660 to 1720, an interval which corresponds fairly closely with the period when European sunspot records were un-usually scarce.

Interesting as are these attempts to connect sunspots with the weather, we are at once on firmer ground when we turn to look at the sun's relationship with magnetic disturbances on the earth. Here the evidence is definite and decisive. The need for caution is gone, and to remove any lingering doubts there are two separate lines of evidence which both point in the same direction. There is, in the first place, a regular daily change in the direction shown by every compass needle, the extent of which moves in rhythm with the solar cycle. Every accurate compass needle can be observed, in the course of 24 hours, to oscillate slowly about its average position. A little after dawn the north pole of the needle is at its farthest to the east. During the morning it swings slowly to the west, reaching the limit of its swing a little after mid-day, and during the afternoon and night returns, rather more slowly, to the point from which we watched it start. It is not a big change, but precise measurements show that it is definitely greatest when there are more sunspots.

At other times a compass needle will begin to swing violently to and fro, only returning to rest when the sudden disturbance in the earth's magnetic forces which it is recording has passed away. This is what is meant by a magnetic storm, and it has been noticed that they are most common when there are most sunspots. Both small and large magnetic storms show a connection with happenings on the sun—but in different ways. The largest storms take place, on the average, at times when there are larger spots near the centre of the sun's disc. With small magnetic storms this direct connection is lacking but, as if to make up for it, there is a tendency for these smaller storms to recur at intervals of 27 days, the time which the sun takes, turning on its axis, to present the same face to the earth.

Taken together, these observations provide a definite clue to what is happening. It looks as if magnetic disturbances on the earth are far-off echoes of some other disturbance near the middle of that part of the sun's surface which at the time is turned towards the earth. If the solar disturbance is sufficiently violent it will produce both a big sunspot and a big magnetic storm. But a weaker disturbance may produce a smaller magnetic storm although no sunspot appears. In other words the sunspot cycle, being an indication of solar activity, is also an indication of the extent of related disturbances on the earth. But it is not always possible to blame individual sunspots for what happens. Sunspots, in fact, are symptoms, not causes of the disease.

Records of the aurora of Arctic and Antarctic regions also tell the same story. Like magnetic storms, the aurora has a way of staging repeat performances at 27-day intervals, corresponding with the sun's rotation as seen from the earth, while it is also most frequent when there are most sunspots. This is by no means all the evidence for connecting sunspots with happenings on the earth. But it is sufficient, at any rate, to suggest that wireless is not likely to be immune. As the radio mirrors, on which reception depends, are high up in the earth's atmosphere we should naturally expect that they would be the first part of the earth to be affected by any influence arriving from outside. The only difficulty is, that radio being a comparatively recent invention, we can no longer talk in terms of complete cycles.

However the sun's influence is exerted, we may reasonably suppose that its effect is to strengthen the radio mirrors, making them capable of reflecting shorter wave-lengths when there are many sunspots than when there are few. Seeing that the effect of sunspot maximum is to increase two other forms of electrical or magnetic activity on the earth (magnetic storms and the aurora), it is quite in order that the activity of the radio mirrors should also be increased. That there is a very definite effect is illustrated by both amateur and commercial experience, although accurate scientific records covering even one complete cycle are not yet available.

An American scientist, Dr Harlan T. Stetson, has specialised in this branch of radio research since 1928. What he does is to measure the exact signal strength with which standard broadcast transmissions from Chicago are received at a number of different stations.[1] Fig. 16 below represents some of his earliest records. It shows the way in which radio reception on a medium

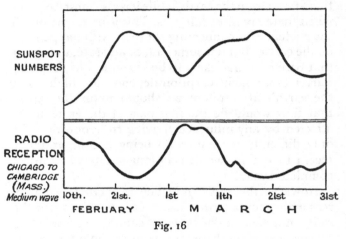

Fig. 16

wave-length responded to sunspot variations during the months of February and March, 1928. This, it will be remembered, was a year of sunspot maximum.

This particular curve is interesting as showing how closely radio conditions may follow changes on the sun. It may also be noticed that the two periods of

[1] Dr Stetson was preceded in this work by Dr G. W. Pickard who began his records at a private laboratory at Newton Centre, Massachusetts, in 1926. Dr Pickard's records are now included in a larger scheme sponsored by Dr Stetson.

worst reception are separated by the now familiar interval of 27 days. But for longer records we must turn to other sources, one of the most striking being the experience of commercial firms operating short-wave services across the North Atlantic. It may be noticed that the effect of the sunspot cycle was in this case in the opposite direction, an apparent contradiction which will be explained later. The point, for the moment, is that a definite effect of some kind was produced.

In accordance with the solar cycle, the years from 1926 to 1928 saw a steady increase in the strength of the radio mirrors, and wireless engineers finding themselves able to use shorter and shorter wave-lengths believed that they were making genuine progress. The "short-wave renaissance" was in full swing and, as with all new movements, hope swung higher than time has justified. The day came when short-wave radio became more difficult instead of easier and it gradually became evident that the sun was to blame. At the peak of sunspot activity a wave-length of 13 metres was in one case being used with complete satisfaction to cross the North Atlantic. By 1934 the transmitters of the same company had been pushed up to 70–90 metres.

From this point of view it is interesting to notice that the B.B.C.'s Empire services were started at very nearly the worst possible time. All these short-wave services would probably be improved by having stronger radio mirrors available. But the difficulties of 1933–4 were particularly noticeable on the Canadian service, where in winter a large part of the wireless

6-2

waves' journey must necessarily be performed in darkness when the radio mirrors are again at their weakest. The B.B.C. started with a wave-length of 49·6 metres, about the longest "short" wave-length available for broadcasting, but it was frankly admitted that a wavelength of 70 metres was really necessary. Perhaps it may be some consolation to the corporation's technical experts to have got over the worst at the start, while disgruntled listeners in Canada can reflect that shortwave radio conditions are bound to improve steadily during the next few years. In this case the lesson of the sun's cycle is obvious and has, indeed, already been learnt. Wireless engineers will always want to use the shortest wave-length which is convenient, because the farther they go down the farther also will they get away from atmospherics. But as the shortest wavelength which can get through on any particular route is always changing, it will be necessary to have a number of different wave-lengths available which can be brought into action at different stages in the 11-year cycle. At the same time there are probably more atmospherics at sunspot maximum, when the lowest wave-lengths can be used. So that all-round conditions are probably best somewhere in between—i.e. in the present cycle about 1936 and again about 1940.

Unfortunately the effects of the sunspot cycle on different classes of wave-length are complicated. It may be helpful to remember that sunspot minimum, winter and night, by reducing solar activity on the earth, each change the condition of the radio mirrors in the same direction, while the opposite changes are produced by sunspot maximum, summer and daylight.

The most marked effect noticed by amateurs is in the reception of medium-wave American stations. For medium wave-lengths the radio mirrors will always be strong enough to provide reflection. The only question, therefore, is to what extent the waves are absorbed in the process. On this point theory and practical experience are in complete agreement. Absorption is greatest when the mirrors are strongest. The reason, it may be remembered, is that the electrons of the reflecting layer are then most likely to lose the energy of their "welcoming dance" before passing it out again as the reflected wave.

The difference in Transatlantic reception has been very noticeable. In the early hours of the morning, round about 1923, when sunspots were fewest, American medium-wave stations were coming through excellently and amateurs in England took pride in receiving them. Then, as the solar cycle wore on, the dance bands of New York and Chicago no longer invaded the peace of European nights. And not until the winter of 1932-3, when sunspots were again at their lowest, could jazz enthusiasts obtain their longed-for extension of listening hours. For the same reason the present is also a good time for European medium-wave stations although, since we can generally hear them anyway, the effect is not so marked. Nor need there be any qualification in this case about atmospherics, since these should be at their most infrequent[1] at the same time as signal strength is strongest.

Long waves are differently affected again. For the

---

[1] I.e. in so far as there are fewer thunderstorms over the world as a whole at sunspot minimum than at sunspot maximum.

ordinary listener, tuning-in to Daventry or one of the high-powered European stations, atmospherics are probably the most important factor. The spot-less period now passing has therefore been the best. But for long-distance transmissions the position is complicated by the fact that for strong reception long waves prefer the Heaviside layer to have a sharply defined lower edge. This is most likely to happen when the sun's activity is greatest. There is thus the paradox that long waves, like short waves, on the whole prefer sunspot maximum for long-distance transmissions, while sunspot minimum is unquestionably best for medium wave-lengths.

We have now seen how happenings on the sun are continually affecting radio conditions. How, it is natural to ask, does the sun do it? In the first place it is quite clear, from what astronomers can see of the sun's surface, that sunspots are not an isolated phenomenon but merely the most obvious sign of more general activity. For example, at times when there are most sunspots, great clouds of gas high up in the sun's atmosphere are also more prominent. Again, watching the edge of the sun, astronomers can see an unusually large number of solar "prominences", great jets of red flame reaching thousands of miles out into space. The technique is completely different in each case, but the conclusions to be drawn from the three sets of observations are very much the same.

Finally, during total eclipse, when the main disc of the sun is entirely covered by the moon, it is possible to photograph even longer luminous streamers. These streamers, which in a sense may be regarded as the

extreme limits of the sun's atmosphere, may be longer
than the sun's diameter. Their briefly seen splendour
during the minutes of totality is a sight not readily
forgotten. But anyone who has seen more than one
total eclipse of the sun will remember that the form
of the corona is not necessarily repeated on different
occasions. Successive observations have shown that
the pattern which these luminous streamers adopt
depends on the sunspot cycle. When sunspots are few,
the streamers are folded into something not unlike a
pair of wings stretching outwards on either side of the
equator. But at maximum activity the corona resembles rather a flower with long thin petals stretching
out to equal distances in all directions. It seems therefore that whatever power, deep within the sun, is
responsible for the great surface vortices which we call
sunspots, it also controls the shape in which the outer-
most regions of the sun reach out to greet the earth and
planets. Nor is it difficult to believe that from the
slowly changing tips of flower or wing feathers some
purely physical influence is passing outwards. If this
be so, it is clear that both the direction and strength of
this radiation will depend on the form of the streamers
from which it comes. So that the sun's connection
with aurorae, wireless and weather, if still a mystery,
is no longer to be numbered among things incom-
prehensible. The sequence is suggestive although the
cause is obscure—first the inner disturbance, then the
vortices on the sun's visible surface, next the appearance of denser clouds of vapour and long tongues of
flame reaching thousands of miles high, and finally
the changing streamers, so natural a route for part of

the sun's radiation to follow on its outwards journey into space.

How does the sun do it? What is this radiation which, strongest at times of magnetic storms, produces such important changes in wireless conditions on the earth? Is there only one kind of radiation? Or is there a normal radiation which is responsible for the day-and-night and winter-and-summer movements of the radio mirrors; and an abnormal radiation, which by its presence or absence is responsible for the more majestic sequence of the 11-year cycle? These, certainly, are questions of mainly academic importance. Yet behind them lies the whole riddle of long-distance radio. The attempt to answer them has taken scientists half across the world to watch two total eclipses of the sun, and up into the Arctic to make special records in connection with the International Polar Year of 1932–3. Just because the questions asked are so difficult, the ways in which they have been answered are of unusual interest.

What the sun has got to do, it will be remembered, is to knock spots off the particles of the upper atmosphere, the spots being the electrons of the Heaviside and Appleton layers. There are several ways in which the particles of a gas can be maltreated in this fashion. One way, and this can easily be done in the laboratory, is by bombarding it with X-rays. But above X-rays in the long sequence stretching from the "gamma" rays of radium to wireless come ultra-violet rays. These are included in the sun's radiation, and doctors have lately begun to take great interest in them, regretting that a large proportion of ultra-violet rays are absorbed during their passage through the misty atmosphere of winter and sometimes going so far as to attribute a certain amount of winter ill-health to their absence. Ultra-violet rays must, therefore, at the best of times be very much stronger near the top of our atmosphere. So, as they also are capable of knocking spots off a gas, they must clearly be regarded as strong candidates for the honour of maintaining our radio mirrors.

Experiments in the laboratory also provide the clue to the second possible way in which the sun may, and in fact does, affect wireless conditions. A whole variety of minute particles, such as scientists obtain when they split atoms, can also knock spots off the particles of a gas. This is a direct case of bombardment in the ordinary sense of the word, although the bombardment is not so strenuous as that needed to split the inner part of an atom. We may notice, however, in passing that "ionisation" produced in this way is of the greatest practical use to scientists bent on exploring the atom. Ionised particles, or particles which have had spots knocked off them, have a way of acting as centres for the condensation of water vapour into droplets. So when Lord Rutherford shows a photograph of the track of some particle which he has succeeded in hurling out from inside an atom, his audience is really seeing the thin trail of fog which the fast-moving particle has left behind as a memento of its passage through water-laden air.

Records of this kind do at least suggest that particles hurled out from the sun may have some share in keeping the earth's radio mirrors active; and the same idea is encouraged by those eclipse photographs, to which I have already referred, which show luminous streamers reaching far out from the sun's surface into space. The sun certainly looks, during eclipse, as if something tangible was passing out along its streamers. But while the something tangible must presumably consist of particles, there is no clue from eclipse records as to whether these particles are electrically charged or are "neutral", that is carry neither a positive nor a negative charge. This question affects very definitely the way in which we should expect them to behave as they approach the earth.

The earth, as we know, is a large magnet and magnets not only have an attraction for iron, but can also deflect electric currents. Now an electric current is only moving electricity, and a stream of charged particles clearly comes under the latter definition. We should therefore expect that they should be deflected when they come within range of the earth's magnetic forces. It is possible to make a small model of the earth complete with north and south poles and to show how incoming charged particles converge upon the Polar regions in a series of narrowing spirals; and the Norwegian scientist, Professor Størmer, has calculated that the whirling track of such particles as they approach the neighbourhood of the North Pole corresponds quite closely with many of the observed shapes of the Northern Lights.

A stream of electrically charged particles therefore accounts perfectly for the appearance of the aurora in Arctic and Antarctic and for its absence from more temperate regions. But it is not of much help in connection with wireless because there are perfectly good radio mirrors over England, and for that matter over the equator, as well as at the Poles. So what we may call the particle theory of the sun's effect on wireless was put forward in the form that there were "neutral" as well as electrically charged particles and that, while the electrically charged particles gravitated to the Poles, the neutral particles were responsible for keeping the radio mirrors active over the earth as a whole.

We can now see how the total eclipses of 1927 and 1932 come into the story. They provided a test of the speed of the effective radiation, and they did this because the essence of a solar eclipse is that the moon blocks the path of whatever is coming from the sun to the earth. How and when this will happen clearly depends on the speed at which the radiation to be cut off is moving, just as an anti-aircraft gunner is in effect allowing for the limited speed of his projectile when he aims in front of a moving target. In the case of the sun's bombardment, ultra-violet rays travel, like light waves, at a speed of 186,000 miles a second, while it is believed that particles thrown out by the sun travel at the relatively modest speed of about 1000 miles a second. The latter figure is based on the calculations of the British astronomer, Professor E. A. Milne. But observations made at Greenwich Observatory suggest that it is probably not very far wide of the mark. It has been already mentioned that apparently direct connections have been noticed in some cases between the appearance of particular sunspots and particular magnetic storms on the earth. For the larger storms, at any rate, the time interval between the passage of the spot across the centre of the sun's disc and the advent of the magnetic storm seems to be about 1 or 1$\frac{1}{2}$ days.[1] On this basis it can be calculated that the radiation responsible for the disturbance has travelled the 93,000,000 odd miles from the sun at an average speed of between 700 and 1100 miles a second.

It follows that whenever the sun's light is eclipsed, there will also be a "particle eclipse", although not quite at the same time and place. As the particles travel more slowly than light we might at first sight imagine that the particle eclipse would follow the ordinary light eclipse, which so many people travelled to the north of England to watch in the early morning of June 29, 1927. Actually the problem is more complicated than that and, when the motions of the moon and earth are taken into account,the answer works out that the particle eclipse should happen appreciably earlier than the light eclipse, and that the area of total eclipse should be shifted towards the east. The necessary calculations have been made by astronomers, and they are no doubt right. They say that the difference must have been about 1$\frac{1}{2}$–2 hours both for the English total eclipse of 1927 and for the American eclipse of 1932.

The first evidence in favour of the theory that ultra-violet rays are responsible for maintaining the strength

[1] Examination of the records in the case of smaller magnetic storms suggests a somewhat longer time interval, and other workers have suggested intervals ranging up to 4–5 days. The latter figure would correspond with a radiation speed of rather less than 300 miles a second. So long as there were no particles of very much higher speed than seems likely the particle eclipse referred to in the following paragraph would still be adequately separated from the light eclipse. The effect of other particles of lower speeds would be to prolong the duration of the particle eclipse.

of the radio mirrors was obtained by English scientists during the 1927 eclipse. To help them make the most of their opportunity special transmissions were provided by the B.B.C. from Birmingham, Newcastle, London and Manchester. Ships at sea were also asked by the Post Office to use their transmitters as little as possible around the time of eclipse in order that there should be no undue interference. A number of useful records were obtained by both professional and amateur observers in different parts of the country, but the most convincing were those of Professor Appleton at Peterborough where he received the Birmingham transmissions. He found that the height of the lower of the two radio mirrors jumped up 20 miles during the time of ordinary eclipse, and that the intensity of waves reflected from it was increased more than ten-fold. Moreover, the effect was at its greatest within a few minutes of mid-eclipse, compared with the hour or two's difference which would have been demanded had the suggested particle eclipse been responsible. Farther north, an amateur observer at Giggleswick, listening-in to the special London transmission, noticed almost equally striking changes.

Here then was direct proof, so far as the lower of the two radio mirrors was concerned, that the sudden cutting off of the sun's ultra-violet rays might drastically affect wireless conditions. The only difficulty was that it was not at the same time possible to say that the cutting off, an hour or two earlier, of the sun's particle stream had not also made itself felt on wireless reception. Although the Peterborough records had been begun at 2 a.m., more than two hours before total eclipse, the hours covered were those of the usual dawn changes in the radio mirrors when it was not possible to say that the effect of the supposed "particle eclipse" might not have been missed. Moreover, although Professor Appleton professed himself tentatively satisfied with the result of the test, there were many who felt that the radio changes during this eclipse might have represented nothing more exciting than a lucky coincidence.

So when the time arrived for the 1932 eclipse, very thorough arrangements were made by the Canadian National Research Council to make the test as complete as possible. A Canadian scientist, Dr J. T. Henderson, even came over to London to be fully trained under Professor Appleton in the methods to be used, returning to Canada in time for the eclipse. As things turned out, he made his reputation as a radio scientist by the thoroughness with which he played his part on this occasion. The test was also helped by the fact that the eclipse took place at a convenient hour in the afternoon, when no abnormal changes would be expected to disturb the radio mirrors.

Three separate stations were established, each with its own part to play in this radio detective drama. Dr Henderson's own station at Vanleek Hill, Ontario, was chosen so as to be as nearly as possible immediately below where the central line of the light eclipse would pass through the Heaviside layer, 70 miles up. Here he was able to show, beyond possibility of criticism, that there was a 58 per cent. decrease in the strength of the lower radio mirror and that this decrease was at its greatest within a few minutes of mid-eclipse. Meantime D. C. Rose of the National Research Laboratory, Ottawa, working at the second eclipse station at Kingston, had found that there was a similar decrease of roughly half this amount in the strength of the upper radio mirror. This also took place at the time of the ordinary light eclipse at that height. So in each case there was good reason for saying that the cutting off of the sun's ultra-violet rays was responsible. The part of the third station at Corner Brook, Newfoundland, was to look after the supposed particle eclipse, the central area of which unfortunately fell over the middle of the Atlantic. Corner Brook was, however, some 200 miles inside its western boundary. This was not an ideal situation, but it was the best that could be secured, although the difficulty of local conditions prevented anything like complete records being made. Such records as were obtained gave no evidence of the particle eclipse. They therefore, so far as they went, confirmed the opinion of the other two stations that ultra-violet rays were the part of the sun's rays which chiefly affected the radio mirrors.

The whole question was cleared up by one final expedition to the Arctic which, in this phase of radio research, may be said to have ended the period of adventure. The occasion was the International Polar Year of 1932–3, when expeditions from 12 countries set out with an agreed programme of research to make every kind of record which it was felt could usefully be kept in Arctic regions. Since a great deal of the northern hemisphere's weather comes, directly or indirectly, from the Arctic, the obtaining of meteorological records naturally bulked large in the plans of many of the expeditions. Others, by making measurements of the aurora borealis, contributed indirectly to radio science. But one of the two British parties, which went to Tromsö on the north coast of Norway, was solely concerned with wireless problems. The whole venture was the jubilee of the first international polar year of 1882–3. But many more expeditions took part and the field of inquiry was greatly widened, the inclusion of radio being an obvious addition to the programme of 50 years earlier.

It is easy to see why this expedition to an eminently civilised part of the Arctic was able to prove finally that ultra-violet light from the sun is responsible, at any rate, for the workaday efficiency of the radio mirrors. Working in a high latitude no very elaborate records were needed to settle the matter once and for all. It was simply a question of whether at the equinox, when day and night are of equal length, the radio mirrors were stronger or weaker than in England. It is a test that follows quite directly and straightfor-wardly from the geometry of the earth and sun. The farther north we go, the lower in the sky does the sun appear owing to the fact that the sun is, so to speak, opposite the equator. This is, of course, only an extension of the well-known fact that at noon on the equator the sun is directly overhead, while even at midsummer it never attains to that position in England. It follows that as we move more and more to the north, the sun's light (including its ultra-violet rays) arrives on a more and more slanting course, having travelled farther and farther through the atmosphere. More ultra-violet rays will therefore be absorbed *en route* and, if they are the agency which keeps the radio mirrors active, then the radio mirrors will be to that extent weaker. If, on the other hand, the agency responsible is a mixed bag of charged and uncharged particles, then the charged particles will be deflected in a narrowing spiral towards the two poles of the earth and, on balance, we should expect the radio mirrors to be stronger.

The Tromsö party, again under Professor Appleton, was able to settle the question very neatly. They found that the radio mirrors, although broadly similar to those over England, were 30 to 40 per cent. weaker at the equinox. So far, so good. Ultra-violet rays were responsible. But this greater weakness of the radio mirrors was only found on days when wireless conditions were steady and well-behaved. On other days phenomena were observed which go a long way towards explaining why radio communication in polar regions will always be liable to interruption. Often there would be magnetic storms and displays of aurora borealis, both the latter suggesting the arrival of charged particles from the sun, and on these days radio conditions would be altogether different. On days when mild magnetic storms took place, generally at night, the radio mirrors behaved in the way expected. Their strength increased, even although no ultra-violet rays from the sun could possibly be reaching them, and it was evident that charged particles from the sun must be responsible.

But with intense magnetic storms something completely unexpected happened. The radio mirrors first of all increased in strength but, as magnetic conditions became more disturbed, were driven completely out of action. The only explanation was that at these times the electron network became so dense and reached so far down that wireless waves, instead of being reflected, were completely absorbed. It is an anomaly which at once explains the difficulties of communication in polar regions. Moreover, any doubt that these periods of blank reception were connected with the sun was removed by the fact that they tended to recur at 27-day intervals—the length of time that the sun, and particularly sunspots, take to go round.

It is now clear why all touch was twice lost with General Nobile in the course of his polar flight in the airship *Italia*. The two periods of silence, which naturally caused considerable anxiety, were about 27 days apart; and in the light of the polar year records there can be little doubt that both periods of silence were times of intense magnetic activity, caused by the appearance on the sun's surface of the same sunspots. It follows that periods of blank reception in the Arctic and Antarctic will always be most frequent at times of sunspot maximum. Future expeditions which intend to depend at all largely on radio communication will therefore do well to select, unlike General Nobile, a time when sunspots are few. At the same time, as the sunspot cycle was already approaching minimum during the polar year, it is obvious that polar explorers will never again be able to put over-much faith in their wireless transmitters.

From the point of view of the scientists, however, the radio records obtained during the polar year are eminently satisfactory. The apostles of both theories have been justified. To ultra-violet rays must be given the credit for keeping the radio mirrors in their normal active state. To electrically charged particles must be assigned the less creditable rôle of introducing abnormal disturbances, along with magnetic storms and the splendour of the aurora. Also, since electrically charged particles tend to congregate near the poles, it is clear that these abnormal disturbances will always be greater on radio routes that either pass far to the north or the south.

There is one other kind of radio disturbance which can be conveniently described at this point, that caused by thunderstorms. Thunderstorms are responsible for a large proportion of atmospherics,[1] for most of us a much more practical problem than any difficulty which may arise in polar communication; and their inclusion in a chapter nominally devoted to the sun may be justified by the possible connection, already mentioned, between sunspots and thunderstorms.

Although no one will be inclined to doubt the power behind a lightning flash, some records made by Professor C. T. R. Wilson at Cambridge may give a more accurate idea of how great it really is. Professor Wilson has played a bigger part than anyone else in bringing the thunder cloud and lightning flash within the scope of scientific measurement. Some of his conclusions are, to say the least, impressive. Among other things

[1] The fact that other weather conditions may also give rise to atmospherics is discussed in Chapter XI, "Radio and the Weather Fore caster".

he finds that a cloud which is giving flashes at the rate of three a minute is using up electrical energy more than a thousand times as fast as does Europe's most powerful broadcasting station. Or, bringing the comparison nearer home, we may say that a single flash would be capable, if fully translated into heat, of providing the hot water for a thousand good-sized hot baths.

An English weather expert has taken the matter a step farther. He has calculated that over the world as a whole there must be an average of about 44,000 thunderstorms a day. So that at any given moment we may imagine that 2000 separate storms are engaged in banging and flashing away their energy in different parts of the world. On this basis the earth provides sanctuary for perhaps a hundred lightning flashes a second and the power output which they represent is about 2000 times as great as the sum of the official ratings of Europe's 14 most powerful broadcasting stations. Equally, if the whole of the world's thunderstorms could be concentrated over London and all their energy utilised for the purpose, they could be made to provide a hot bath for every man, woman and child in Greater London in a little over a minute.

More than one theory has been put forward to explain the origin of these enormous electrical forces. But of their magnitude there is no doubt at all. It is not, therefore, surprising that thunderstorms can be a nuisance. Indeed, if all their energy were transformed into the radio waves we know as atmospherics, it is not too much to say that any form of wireless communication on ordinary broadcasting wave-lengths would be impossible. Fortunately by far the greater proportion of the energy of a lightning flash is devoted to warming the air on its journey from cloud to earth. But, although the sending out of radio signals is quite a side-line from the point of view of a thunder cloud, the signals are strong enough to be picked up and identified by special radio receivers 4000 or more miles away. In a later chapter we shall see that this raises interesting possibilities in the way of weather forecasting. But for the moment we are more concerned with the effect of atmospherics on radio reception, and this must obviously depend on the wave-lengths of these most unwelcome signals.

It is possible both to photograph the wave-shape of individual atmospherics and to measure how long each disturbance lasts. This has been done in England by the Radio Research Board. Records have also been made on an experimental cruise to Port Said and the Bay of Bengal, as well as at Helwan near Cairo and at Khartoum. More recently the Union Radio Scientifique Internationale has made arrangements for the setting up of a general world network of atmospheric-recording stations. The general conclusion is that, while the forms of individual atmospherics may be both varied and complicated, the time which each individual "pulse" lasts is of the order of a thousandth of a second. That means that nature operates on a wave-length of about 300,000 metres, which at first sight might seem to be far enough removed from the wave-lengths of ordinary broadcasting to secure immunity from interference.

The trouble is that a lightning flash is, from the

point of view of the radio engineer, a singularly poorly controlled type of transmitter. It sends out, not a chain of waves, but an individual wave and that of complicated and irregular form. So the term wavelength, when applied to atmospherics, has little definite meaning. It is probably true that if it were physically possible to build radio transmitters big enough to send out 300,000 metre waves, we should find for our pains that atmospheric interference was at its greatest somewhere about that wave-length. But, as we all know from experience, the effective wavelengths of atmospherics reach well down into those of broadcasting; and the only certainty is that the farther we get away from their primary wave-length the less interference shall we suffer from that source. There is nothing, therefore, to be done about atmospherics. They are there and we must put up with them. The only consolation is that wave-lengths below about 10 metres are virtually immune from their unwanted cacophony; and this, as we shall see in the next chapter, is one of the great advantages of ultra-short and micro waves.

CHAPTER VI

THE FUTURE
AND ULTRA-SHORT WAVES

We have reached the end of the long story of the radio wave. We have seen how it comes to be produced, how it is like light and yet different, and how it gets round the world. We have seen, too, how its power waxes and wanes in daytime and at night, in winter and summer, and with the sun's slow cycle of changing activity. All these things, it has been learnt by experience, are of practical importance. But we want to know the future of radio as well as its present; and, if scientists have rightly read the riddles of the strange world above them, it should not be impossible to obtain a glimpse or two of what wonders are still in store.

In some ways the vision may be disappointing. Man can improve the efficiency of transmitters and receiving sets, extend the range of wave-lengths which he can use, and make better use of the weapons at his disposal by always choosing the right wave-length for the right job. But there his power ends. He cannot hope to control the wireless mirrors of the sky or to abolish atmospherics. The mirrors on which he must for ever depend for distant reception are beyond his reach, and thunderstorms are as little amenable to royal commands as are the tides.

For what I may call domestic broadcasting, by which I mean broadcasting within a single continent, the limits of progress may be very simply laid down. The greatest scope for improvement is in the receiving set, both in faithfulness of reproduction and in power of selection. For the rest transmission is almost perfect; the ether is grossly overcrowded, a misfortune which must ultimately be settled by common-sense negotiation between the powers; and day reception must inevitably remain better than night, a limitation of no very serious importance. We can get our own stations at any time when there is a programme available, and it is not as a rule until after dinner that we begin to think of sampling the free entertainment provided by other countries.

There remains the great problem of fading. This, as we have seen, may be caused by interference between the ground wave and the wave which has been reflected at one or other of the radio mirrors. Another cause of fading may be the fact that wireless waves are scattered, rather than reflected in the strict sense of the word, from whichever of the radio mirrors happens to be in action at the moment. The area from which reflected waves may reach the receiver may be as much as 10 miles wide, so that there is appreciable room here for interference. In any case it is the mechanism of reflection that is at fault, and that is something which we can never hope to control. Some improvement might be effected if a way was ever found of cutting out, in the case of distant stations, the relatively faint wave which has come along the ground. But it must be admitted that at present there is no obvious way in which the offending wave could be removed. So with this very bare sketch, and the promise that television will be discussed in a later chapter, we must pass from domestic broadcasting to wider fields.

In long-distance broadcasting, and particularly the B.B.C.'s Empire services, there is very much more room for improvement. It is here that the engineer is most seriously worried by the vagaries of the radio mirrors and by the changes of the sun's cycle. Broadcasts which travel half-way round the world must bounce many times between earth and sky. In most cases it is difficult to avoid the changing periods of sunset or sunrise at some part or other of the route, and so the mirrors will be always liable to move and there will always be some possibility of fading.

But with more experience in choosing the best hours for broadcasting and the best wave-lengths to use, it should certainly be possible to reduce this difficulty. Equally, there is room for improvement in the design of the aerials at Daventry. Transmission can be made either more or less directional, and only trial and error, backed up by detailed reports from the whole areas to be served, can show which are the best arrangements.

More important still, it will not be possible, until the whole 11 years of the sunspot cycle are over, for the B.B.C. to tell what are the most suitable wave-lengths at all periods of the cycle. Fortunately for the reputation of their service the most difficult possible time has been chosen to make a start. The year 1933 saw the lowest ebb of the cycle and the radio mirrors correspondingly at their weakest. Under these conditions and in winter, when the hours of darkness are long in northern latitudes, it is admitted that a wave-length as long as 70 metres is really desirable on the Canadian route.

At present, international agreement forbids the use of any wave-length between 50 and 200 metres for broadcasting purposes. It is hoped that this restriction on a band of long-distance, and therefore essentially useful, wave-lengths will be modified before the next period of difficult conditions arrives. It is certainly unfortunate that wave-lengths which fill an essential gap in world broadcasting should be monopolised by ships or fighting services.

In the meantime conditions on all the Empire routes will steadily improve until somewhere about 1938 when sunspot maximum will again bring the turn of the tide. As the years go by it will be possible to use shorter and shorter wave-lengths, and to get farther and farther away from atmospherics. Nature's preference for long wave-lengths is a factor which must be of the greatest importance in Empire broadcasts. It means that it will always be in the interests of engineers to use the shortest wave-length which they can count on getting through. In any permanent service it will be necessary to have a number of different wave-lengths available for use in summer and winter at different stages of the cycle.

Much the same considerations apply to long-distance telephony, in the development of which the British Post Office has played such an important part. It is now possible to speak from Rugby direct to Canada, the United States, India, South Africa, Australia and Egypt. Very soon a service to China, and thence to Japan, will also be available, and even then it must be admitted that only the fringe of the possibilities of this form of communication will have been touched. Unless the whole trend of world trade is reversed business men must inevitably feel more and more the need of direct contact with their colleagues in other countries, and the growing tendency towards international discussions of all kinds must have the same result.

The Post Office makes use of short-wave transmitters on all these services, but there is also at Rugby one of the longest wave transmitters in the world.[1] This can reach any continent under practically any conditions and can therefore be conveniently used to fill in any gaps in short-wave reception. It is as if a burglar were to blow up a whole building in order to open a safe because in a difficult case he had not sufficient experience of his craft to use the proper tools. With the growth in long-distance telephony it is obvious that more and more effort will have to be directed towards securing 24-hour reliability on short wave-lengths, or as near that as may be. This again must mean the multiplication of wave-lengths and the knowledge to direct their use. The burglar must extend his kitbag, and learn what each tool can do.

Within the last few years the use of "ultra-short" and "micro" waves has opened up a whole new range of practical radio. The possibilities range from television of real entertainment value to the treatment of disease. But as this sort of wave has so far come hardly at all into the story we shall have to make quite a long digression before we can come to the future.

[1] The longest wave station in the world is Sepetiba, Brazil, 21,820 metres. The wave-length of Rugby is 18,750 metres.

"Ultra-short" is generally taken to mean any wave-length too short to be ever reflected by either of the radio mirrors to which the greater part of this book has been devoted.[1] That is an obvious and entirely practical distinction. The term "micro" wave is a little more puzzling. It might be thought that no wave could be shorter than ultra-short, just as no petrol is ever claimed to be more than ultra-quick-starting, and no modernist ever claims to be more than ultra-modern. But whatever the logic of the situation the fact remains that when scientists first produced radio waves shorter than one metre they felt that the occasion was sufficiently momentous to demand a new name. So a new name there had to be. At present micro waves mean anything shorter than one metre, but perhaps in time, as wave-lengths are still further reduced, the world may be presented with micro-micro wave-lengths as well.

At the time of writing the shortest wave-length in practical use is 17 centimetres, rather less than one-fifth of a metre. These waves are more than a 100,000 times shorter than those sent out by the Post Office's long-wave transmitter at Rugby. The gap is so enormous that we should expect the two sets of waves to behave quite differently. Wireless wave-lengths, whether long or short, are only part of a very much longer series of wave-lengths, which includes not only light and heat waves, but X-rays and the so-called gamma rays of radium as well. There is not in most cases any hard and fast line. An electric lamp produces heat waves

[1] Anything below 10 metres, according to the International classification previously quoted.

as well as light; there are stars whose "light" is believed to consist mainly of X-rays; and X-rays in turn can mimic the rays naturally produced by radium. So it will not be very surprising if the shortest radio waves are quite as much like heat waves as they are like the waves of broadcasting.

We shall get a better idea of the versatility of "electro-magnetic" waves (for all these kinds of waves are covered by that mysterious term) if we run quickly through the whole gamut. We shall also see how large a place radio occupies in the complete range, and how much we may expect radio waves to differ among themselves. As a yardstick we will use the octave of music, although sound waves are of quite a different type and have no connection with the electro-magnetic family. The octave is, however, a convenient ruler which can be fitted to wireless, or indeed any other kind of wave, just as well. It is an indication, not of the amount of room available for communication, but of the extent to which we may expect the properties of waves to change as we pass down the scale.

The particular harmony, which we call the octave, corresponds to the physical fact that the frequency of the sound waves has been doubled. Frequency, it will be remembered, means the number of waves that are sent out a second—in the case of music, the number of times a second that the string of a violin, or the vocal chords of a singer, move backwards and forwards. So the wire which sounds the middle C of a piano moves only half as quickly as the one which sounds the C an octave higher, and so on right up into the treble.

Altogether about 17 radio octaves are available for

practical use compared with the 7 octaves of a grand piano. So we may say that the modern radio engineer has as many "notes" to play with as would be contained in a piano about 9 feet long. Such an instrument would certainly require at least two pianists, and the players in a quartet would be less cramped in their activities than are those in a duet on a normal piano. On the other hand the players in the extreme bass and the extreme treble would be wasting their energy. Neither they nor anyone else would be able to hear what they were playing since the range of the human ear is only about 9 octaves, 8 octaves less than the range of radio.

The practical radio range is made up, in round figures, as follows: long waves, 4 octaves; medium waves, 4 octaves; and short, ultra-short and micro waves, 3 octaves each. So that events since the short-wave revolution have more than doubled the useful range of radio, while the introduction of ultra-short and micro waves alone has added a third to its length. It would be surprising if the new radio were not in many ways different from the old. So having indicated the extent of the wireless gamut, we must now try to place it in its proper perspective. By looking at the other members of the electro-magnetic family we shall be able to see the sort of ways in which ultra-short and micro waves may be expected to depart from the accepted standards of radio behaviour.

Fig. 17 will probably provide the clearest idea of the relationship between the various classes of waves. Altogether there are some 70 octaves of electro-magnetic waves, of which roughly 43 are claimed by wireless. Of the 26 wireless octaves which are not at present

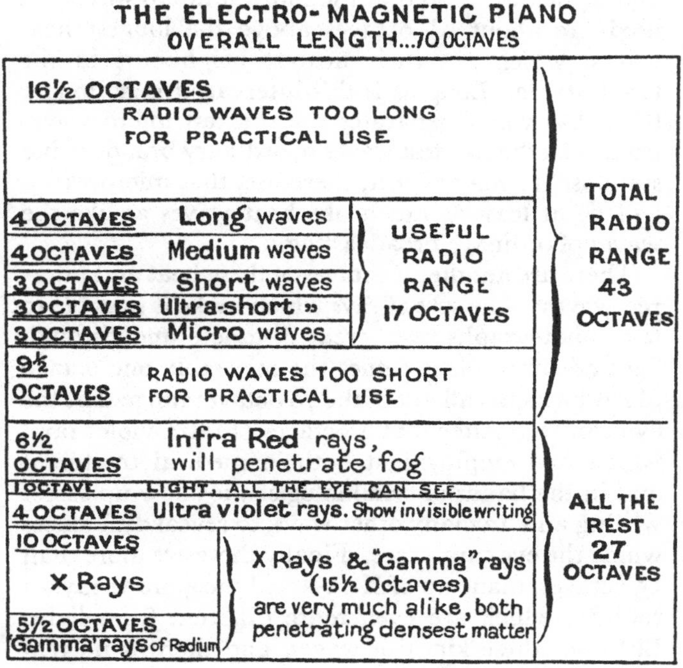

THE ELECTRO-MAGNETIC PIANO
OVERALL LENGTH...70 OCTAVES

16½ OCTAVES RADIO WAVES TOO LONG FOR PRACTICAL USE		**TOTAL RADIO RANGE 43 OCTAVES**
4 OCTAVES Long waves	**USEFUL RADIO RANGE 17 OCTAVES**	
4 OCTAVES Medium waves		
3 OCTAVES Short waves		
3 OCTAVES Ultra-short "		
3 OCTAVES Micro waves		
9½ OCTAVES RADIO WAVES TOO SHORT FOR PRACTICAL USE		
6½ OCTAVES Infra Red rays, will penetrate fog		**ALL THE REST 27 OCTAVES**
1 OCTAVE LIGHT, ALL THE EYE CAN SEE		
4 OCTAVES Ultra violet rays. Show invisible writing		
10 OCTAVES X Rays	X Rays & "Gamma" rays (15½ Octaves) are very much alike, both penetrating densest matter	
5½ OCTAVES Gamma rays of Radium		

Fig. 17. There is an "overlap" of 2½ octaves (not indicated in the diagram) between the shortest radio waves and the longest infra-red waves. There are also similar "overlaps" between infra-red and X-rays, and between X-rays and the gamma rays of radium. This means that wave-lengths within these ranges can be produced in either of two ways. Conversely the absence of any gaps implies that there is no room left in the electro-magnetic scale for any new "mystery" rays.

used, we may say that 16½ are above the practical range and 9½ below it. So that the shortest micro waves yet used are about 9½ octaves above the longest heat waves, which are their nearest neighbours in the family table. Long as is this interval, it is no longer than that which separates the shortest micro waves from even the shortest waves of ordinary broadcasting stations. We may expect, therefore, that micro waves will be at least as much like heat waves as like the waves of ordinary broadcasting.

There are another 9 octaves of these heat and infrared waves[1]—the sort of rays which make it possible to take photographs over great distances and through fog and mist. Next comes the miserable one octave of light which is all out of the 70 octaves range that the eye can see. After that 4 octaves of ultra-violet rays, which find employment both in medical treatment and in the hands of criminologists, to show up secret writing and, in many other ways, to reveal differences which the eye cannot see. Finally there are more than 15 octaves, made up of X-rays and the gamma rays of radium, which are even more different from light. Either of these kinds of waves, and there is no real distinction between them except their source, can be used both to attack cancer cells within the body and for the detection of flaws in armour plate and metal work of all kinds.

It is not really surprising that the eye can see nothing on either side of the reds of sunset and the purple of distant haze. The eye has adapted itself to see just

[1] Overlapping by 2½ octaves with the shortest waves produced electrically (radio).

those wave-lengths in which the sun's rays are richest. As has been already mentioned, other stars are most prolific of other kinds of radiation. No doubt if we had happened to live on a planet belonging to an "ultra-violet" sun our eyes would have learnt to see ultra-violet light and we should have no need of a special lamp to show up invisible ink. And if we had an X-ray sun we should, if we were able to exist at all, in all probability see our fellow mortals in the stark outline of an X-ray photograph. We might even, if we had a suitable sun, be able to see the shortest of radio waves.

What is much more surprising is the way in which man has found practical, even commercial, use for so large a proportion of the total range of wave-lengths which nature has provided. And this brings us back again to ultra-short and micro-wave radio, which from our octave scale we can now see are likely to be about as different from the waves of broadcasting as are the X-rays of medicine and dentists' photographs from the colours of the rainbow.

What are the chief of these differences? In the first place ultra-short waves travel very nearly in straight lines, like those of heat and light and indeed all the other kinds of waves which are members of the electro-magnetic family. This means that transmitter and receiver must be in a straight line, with nothing very substantial in between. Ultra-short waves are only suitable for communication between fixed points, chosen so that the waves have no serious obstacles to surmount, or for broadcasting within a limited area.

The next important difference arises from the fact

that while ordinary radio waves will pass easily through any number of buildings, neither light nor heat waves can penetrate brick and mortar. As might be expected, ultra-short and micro waves are betwixt and between. Marchese Marconi has found that waves of 50 cm. length will get through the walls of an ordinary house. This observation was made in Italy, where houses are mainly of brick and stone. In America it is said that steel framework buildings have proved an effective barrier. That means that for broadcasting in a modern city it would probably be necessary for all receivers to have their aerials on the roof. But that, after all, is not a very serious disadvantage. It would only mean a return in the matter of aerials to the days of early broadcasting.

Before leaving this question of the absorption of waves by matter there is one rather curious point which we may notice. At the long end of the scale broadcasting waves pass through matter with fair completeness; infra-red rays will penetrate mist if not solid matter; and light and ultra-violet rays find almost all kinds of matter opaque to them. Then, at the far end of the scale, X-rays and "gamma" rays will both pass through even the densest matter to some extent so that several inches of lead are necessary to give effective protection against radium.

The explanation takes us back to the atom, which we may remember for the welcoming dance with which its electrons greet an advancing light wave. When talking about reflection I said that this welcoming dance was only kept up for a minute fraction of a second and that its energy was then handed out again as part of the reflected wave. Sometimes, however, the electrons can hold the welcoming dance more or less indefinitely. For any particular atom this can only happen in the case of special wave-lengths. The atom has its own ideas about what kinds of tune its electrons should dance to. All the favoured wave-lengths lie around the middle of the wave-length sequence which stretches from long-wave wireless to the waves of radium. That is why only the waves at either end of the scale can pass through matter.

This is the only case in which our original scale argument from the behaviour of ripples and sea rollers breaks down. In all other ways the character of "electro-magnetic" waves changes continually as we pass down the scale. Even this one exception is not very fair. It is no irregularity in the waves which is responsible for this apparent break in their ordered sequence, but an idiosyncrasy of matter which no analogy taken from water waves could hope to explain.

Most of the other peculiarities of ultra-short waves have already been mentioned at one stage or another. Not being reflected from the radio mirrors of the sky, they are limited to the distance which they can reach along the ground; they are almost entirely free from atmospheric interference, because atmospherics are predominantly of longer wave-lengths; they can be despatched from very much smaller aerials, and, since they can be concentrated into a narrow beam, they require very much less energy to produce the same signal strength at the receiving end.

The shortest wave service in the world, for example, which connects the aerodromes of Lympne and St

Inglevert on either side of the English Channel, only radiates a power roughly equal to that required by a small pocket flash-light. Even if we count the whole of the power used by either station for both transmission and reception, the answer is still only one 250th of the power which will be used in the B.B.C.'s new Droitwich station.

A micro-wave station is also interesting on account of its appearance. The transmitter looks more like a giant searchlight than anything to do with wireless. This is because a solid mirror is used to focus the radio beam just as would be done with light. In practice two mirrors are used, a small one and a big one, and if the visitor looks very carefully between them he will see a small metal rod, about an inch and a half long, with a mushroom top. That is a full-length micro-wave aerial. So, if ever micro-wave broadcasting becomes possible, we need fear no disfigurement of our cities from any further growth in roof aerials.

Altogether there are eight ultra-short-wave radio services in different parts of the world. There is one, for example, between Nice and Corsica, while thousands of telephone users have spoken across the British Post Office's seven-metres radio link between Cardiff and Weston-super-Mare without realising that anything unusual was happening. The remaining six services connect different islands of the Hawaii group. Micro-wave services are at present limited to the cross-Channel link already mentioned and to the Pope's private service between the Vatican and Castel Gandolfo.

These few and scanty applications of the new technique unfortunately give little indication of its future scope. The advent of ultra-short and micro waves has multiplied almost beyond belief the ether's power of running messages for its human masters. Within the short range of 20–40 centimetres there is room for a thousand times as many separate channels of communication as there is in the whole stretch from 200 to 400 metres, which in Europe alone contains some 74 broadcasting wave-lengths. Nothing that has yet been said gives any indication of why this should be so, and to understand the real miracle of micro-wave wireless we must turn for a few moments to the rather difficult subject of what are called "side bands". There will also be this advantage, that we shall learn for the first time why it is often difficult to separate neighbouring broadcasting stations when we are listening-in.

The side-band difficulty arises from the fact that, while radio waves can be transmitted on an absolutely even wave-length as long as they are carrying no messages, they have an unfortunate habit of spreading out on either side of their allotted wave-length as soon as they are asked to do anything useful. To get an idea of how this happens we must go back to nearly the beginning of the story and see how radio waves contrive to carry sound vibrations across space. It is an unfortunate relapse when we have already watched their complicated journeys round the world with such success, but it cannot be helped. Nor need it delay us long.

The picture which we must have in mind is that of a succession of radio waves, always vibrating at the same speed, but their strength periodically waxing and waning in tune with the slower-changing sound waves they are carrying. Here it is on paper (Fig. 18), first a slowly changing sound wave, then a more active radio wave, and finally a radio wave "modulated" so that the form of the sound is impressed upon it.

The last state of the radio wave is obviously more complicated, if also more useful, than the first; and I can only state dogmatically that it can no longer be

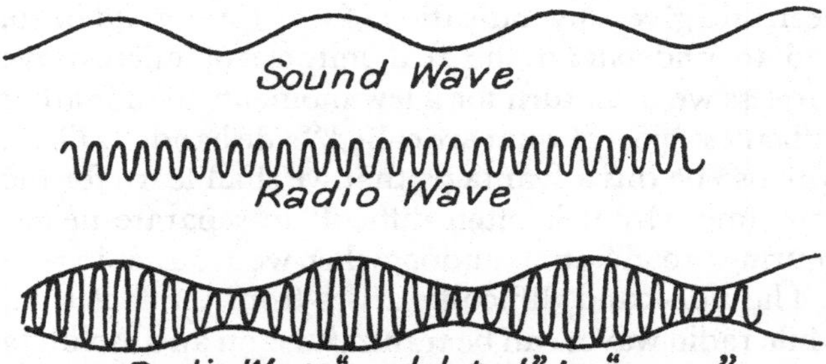

Sound Wave

Radio Wave

Radio Wave "modulated" to "carry"
Sound Wave.

Fig. 18

regarded as a single vibration of constant frequency. As a matter of fact, if we write N for the original radio frequency (number of vibrations a second) and n for the sound frequency impressed upon it, then the result is a complex vibration which may be represented by three separate frequencies—N, the original radio frequency, and on either side of it $(N + n)$ and $(N - n)$. In real life the sound to be transmitted will itself consist of many different frequencies, each of which will add one more pair of frequencies to the modulated wave. How many of these pairs there may be is fortunately no concern of ours. All that we need worry about is that the "spread" of the modulated wave may still be represented by the range $(\mathcal{N} + n)$ to $(\mathcal{N} - n)$, where n now stands for the frequency of the highest sound note which it is desired to transmit. That, therefore, is the extent of the blank interval which we must leave on either side of any broadcasting station if it is not to suffer from interference.

In reckoning up how large this interval should be, we must allow space not only for the highest notes which instrumentalists are going to play, but also for an adequate number of what are called "overtones". These are the related higher notes which accompany every pure musical note and are responsible for the qualities we recognise as tone. Finally when we have decided how many "overtones" we want, and have made room for even the highest of them, there must still be a working margin to allow for imperfections in our receivers. Not even the best of receiving sets can be expected to cover exactly this range at full strength, and then sharply cut off all wave-lengths which are either longer or shorter.

Since the human ear can detect sounds of frequencies ranging up to 10,000 vibrations a second, that must be regarded as the ideal limit for broadcasting purposes. That, of course, assumes the existence of both microphones and receivers capable of rendering these high sound frequencies accurately—a condition which is still far from being realised.[1] But it should none the less be

[1] There is, however, one recently introduced type of receiving set to which this limitation does not apply.

regarded as the ideal. It would be good enough, at any rate, for even the most captious of normal human beings; although it is interesting to notice that a musically inclined dog would, even then, have justifiable ground for criticism since he can hear considerably higher notes than we can.

If we were designing a perfect broadcasting system we should therefore start by insisting that our transmitter be given an effective separation of at least 20,000 cycles (in radio parlance, 20 kilocycles) from its two nearest neighbours in the wave-length scale. As a practical compromise, however, between ideal conditions and the demand of every country for as many broadcasting wave-lengths as possible, an interval of 9 kilocycles has been adopted as the standard separation between stations. This interval is more than twice as small as the least interval which would allow of perfect reception even in theory, the difference being represented by the cutting out of rather more than the highest octave which the human ear can detect. At the same time the ears of most of us are mercifully tolerant of omissions, and the effective spacing between Europe's higher-powered stations is in practice made considerably larger by a judicious distribution of the available wave-lengths.

A glance at any of the published wave-length tables shows that the longer-wave stations are very much more widely separated, in terms of wave-length, than those farther down the scale. We can now see why this is necessary. For Droitwich, with a frequency of 200 kilocycles and a wave-length of 1500 metres, a 9 kilocycle separation involves a very considerable gap in the wave-length table. A frequency of 209 kilocycles means a wave-length of 1435 metres, and one of 191 kilocycles a wave-length of 1571 metres. So that if we are going to stick to our 9 kilocycle separation we can only get three stations within a range of about 140 metres.

When we come down to London Regional, with a wave-length of 342·1 metres and a frequency of 877 kilocycles, the necessary wave-lengths separation on either side is only some 3½ metres; and so on as we progress down the scale. The accompanying table shows the necessary intervals for a wide range of wave-lengths. It will be seen that when we come down to the shortest waves of the radio family the separation

Central Wave-length		Space required for one channel of communication		Number of separate channels for which there is room in space represented by a 10 per cent. change of wave-length
15,000	metres	7110	metres	—
1500	,,	68	,,	2 (in round
150	,,	·68	,,	20) figures
15	,,	·0068	,,	200) in each
1·5	,,	·000068	,,	2000 (case
(15 cms) ·15	,,	·00000068	,,	20,000

Note. The interval allowed between stations in the above is nine kilocycles, the present standard of broadcast separation. Notice that the figures in the last column increase in precisely the same proportion as the wave-length is reduced, i.e. ten-fold in each case.

required becomes almost ludicrously small. For example, micro waves of 15 centimetres would have a frequency of no less than two million kilocycles. So that, if we could tune our receivers sufficiently ac-

curately, a change in wave-length of only nine parts
in two million would be enough to keep the signals of
two stations apart. In fact between 15 and 15·1 centimetres there would be room for more than a thousand
separate stations. Even in the ultra-short wave range
the amount of room is quite large enough to be startling. Turning centimetres into metres, we find that
under ideal conditions ten separate stations could
operate without mutual interference between 15 and
15·1 metres. Actually it is not possible to make either
ultra-short or micro wave transmitters of anything
approaching the wave-length constancy which such
an arrangement would imply. But, as we shall see
later, enough has already been achieved to show how
much the youngest members of the radio family can
do to relieve the congestion of the ether.[1]

As things stand at present, micro waves have just
one big advantage over ultra-short waves. They are
immune from atmospherics, although cars and aeroplanes provide an effective source of interference for
ultra-short waves. The extra shortness of micro waves
is just sufficient to place them beyond interference of
any kind whatever. Against this, there is the drawback that micro waves cannot as yet be produced with

[1] We may notice in passing that, even within broadcasting wave-
lengths, there is one device by which it has proved possible to secure
slightly more room for transmission. This is what is known as the
"single side band" system. It means that the radio wave is only, so to
speak, modulated on one side. With the above notation, the frequency
range is from N to $(N + n)$, the range from N to $(N - n)$ being
"filtered out" before transmission; or *vice versa*. Owing to a certain
amount of extra complication at the receiving end, this system has so
far only been applied to commercial services. But international discussions are proceeding with a view to its adoption in broadcasting.
The result would be to double the amount of space available.

sufficient strength for broadcasting. They can only be used, like a searchlight, in one direction at a time. With these two distinctions in mind we can now go on to consider what the future of both ultra-short and micro waves may be.

The first and most obvious application of these very short waves is as a substitute for submarine cables where relatively short stretches of water have to be crossed. For this purpose the first essential is that a single pair of transmitting stations should be able to handle an appreciable number of conversations at the same time. A modern submarine cable can carry 38 circuits, and this is the sort of figure at which radio engineers must ultimately aim.

Essentially this is simply a problem of wave-length control—and it is at least half-way to solution. On its Cardiff-Weston telephone service the Post Office has just· been able to use twelve different wave-lengths between 4 and 6 metres without mutual interference. Since each speaker must have his own wave-length this represents six telephone circuits, and ultimately it is hoped to increase this number ten-fold.

If ultra-short wave transmissions can be made more strictly directional it will be possible to add still further to the number of separate communication channels. For example, with four transmitters spaced five miles apart and each handling 60 circuits, it would be possible to handle 240 telephone calls at the same time on a fifteen mile length of coastline. Yet this is no idle speculation. It is a perfectly sober statement of what may be expected in a few years' time.

It is also no secret that the Post Office's Bristol

Channel experiment has not been undertaken merely for its own sake. The real object is to demonstrate the possibility of radio-telephone links across the English Channel. A rocky coastline and heavy shipping traffic combine to make the upkeep of submarine cables on this route a costly business. As soon as the reliability of the method has been demonstrated, it is to be expected that the French Government will be approached with a view to a joint service. As telephone requirements develop, additional circuits will be provided by radio and obsolete cables will be replaced as heavy repairs become necessary.

Almost as good results are now being obtained with micro waves. By the time that this appears in print it is probable that a public demonstration will have been given of multi-channel communication on the 17 centimetres link across the English Channel. In that case the balance of advantage would appear to lie slightly in favour of the shorter wave-lengths. The English Channel looks like becoming the Oxford Street of air traffic, and immunity from aeroplane interference may be an advantage well worth securing on this route, just as at the Channel airports now connected in this way it is already almost a necessity.

Keeping for the moment to this strictly limited field, what further possibilities are there for development? Any list of future ultra-short wave links would certainly include New York-Jersey, the north and south islands of New Zealand, the islands of the East and West Indies, Denmark-Sweden, Ceylon-India and Spain-Morocco. Nearer home, we may find England linked to Ireland, the Isle of Man and the Channel Islands in the same way. There are enough possibilities, at any rate, to suggest that there may be a golden future for this, the most modern method of communication.

There are, however, many indications that progress will not stop short at the economical working of these simple and obvious sea routes. Although I said before that ultra-short and micro waves travelled in straight lines, Marchese Marconi has proved that this is not strictly the case. In 1932 Marconi succeeded in establishing communication over a distance of 168 miles on a wave-length of 50 centimetres. The test was made between Roca di Papa near Rome and Cap Figari on the north-east corner of Sardinia. Now Roca di Papa is 2000 feet above sea-level. But even allowing for the heights of transmitting and receiving stations, the curvature of the earth would have prevented reception at distances of more than about 70 miles if the waves had travelled strictly in a straight line.

Marconi's first idea was that his micro waves had somehow been reflected by the atmosphere—although in the light of all the measurements which have been made of the limits of radio reflection it was not very easy to see how this could happen. However, Marconi had done the impossible once, in his original trans-Atlantic experiment, and he was quite prepared to do it again. So far, however, he has been able to obtain no evidence that the waves were coming down at an angle, as would be expected if they had been reflected up in the sky, and a more probable explanation is that not even micro waves travel absolutely accurately in straight lines. We should expect a gradual transition from the wide scattering of long-wave wireless to the virtually straight-line travel of a searchlight, and mathematical theory suggests that this is what happens.

In any case the great importance of Marconi's experiment is that it proves that ultra-short and micro waves can be received at considerably greater distances than simple theory would allow.[1] It also follows that as the power of micro-wave transmitters is increased it should be possible to reach still greater distances. On a practical and commercial basis it is by no means impossible that we may see micro services spanning stretches of water anything from a hundred to two hundred miles in extent.

There is also a further possibility which is of special interest to England as the centre of a wide Empire. The Post Office's Bristol Channel experiment has shown that ultra-short waves can be inserted as a link in a land-line system on a completely automatic basis. Would it be possible to relay a message along a chain of ultra-short-wave stations in the same sort of way that a long land-line provides in effect for a series of distinct relays?

As is generally known relaying on ordinary broadcasting wave-lengths presents considerable difficulties. Most of these difficulties, however, are due to uneven reception. There may be interference from atmospherics or from electrical sources, and the strength of reception on any link may suddenly vary owing to fading. Micro

---

[1] The limit of straight-line travel had previously been exceeded by Mesny and David, working between Nice and Corsica on a 3 metres wave-length, as well as by American workers; but never, either to so great an extent, or on so short a wave-length.

waves are subject to none of these vagaries.[1] For this reason it does not seem at all impossible that speech received at the end of one link should be amplified sufficiently to be broadcast along the next without any serious loss in purity of reproduction.

This prospect raises entirely new ideas in Empire communication. There are many stretches of forest country where the upkeep of land-lines is both difficult and costly. It would only be necessary to select a number of hilltop sites, with more or less uninterrupted views between, and the whole problem of long-distance telephony and telegraphy would be enormously simplified. With, say, five relays per thousand miles, the cost of upkeep would be negligibly small and it would be impossible for disaffected tribes to paralyse communication as can now be done by cutting a land-line at any point.

This, in broad outline, is the promise that ultra-short and micro waves appear to hold for the future of civil communication. It may be taken that it represents, also in broad outline, the ideas of Government experts and advisers on this subject. There are, as might be expected, other lines of probable progress— in television, in medicine, in warfare and in safety at sea. With these I propose to deal in separate chapters. The future of radio, and particularly of very short radio, is too varied to be taken at a single session.

[1] Within, or only slightly over, the range for which a straight-line path is possible.

CHAPTER VII

TELEVISION

THE advent of ultra-short-wave wireless has dramatically transformed the whole outlook for television, and particularly for television in the home, for most of us its most interesting application. So much has, however, been written of the possibilities that would result from transmitting pictures by radio, that little further emphasis is needed. It is enough to remind ourselves that effective screen-thrown television would bring first-class cinema entertainment, as well as real life scenes, to every fireside; that political leaders would be able to exert almost as much personal influence on every "listener" as they can now achieve within the more limited range of a public meeting; and that statesmen, unable to leave their own countries, would be able to participate in international conferences with almost as great realism as if they were present in person. No one doubts the usefulness of television. The essential fact is that it is at long last arriving.

The whole subject of television bristles with technical complications. With these it is not proposed to deal here, but rather to set out the broad principles which it appears likely will govern future developments. The first and most obvious point is that just as telephony depends on the translation of sound into an electric current and back again, so television depends on the similar translation of light. With the transformation of electrical energy into light we are familiar. It is the basis both of the ordinary electric lamp and of the red neon sign.

The opposite translation of light into electricity, which must be the start of television, is something much less ordinary. It is described generally as the photo-electric effect and curiously enough attention was first directed to it by Heinrich Hertz in 1887, the year before he announced the discovery of radio waves. Hertz is therefore in a double sense the grandfather of television. The form which Hertz's photo-electric discovery took was the observation that when ultra-violet light fell on an electric spark gap it increased the ease with which a spark passed across the gap. The explanation, it was found, was that the impact of light on many forms of matter threw out free electrons from it. This is the principle on which the photo-electric cell, or so-called electric eye, works.

A simple form of photo-electric cell, which will serve to make the general arrangement clear, consists essentially of a glass bulb, silvered internally and with a very thin layer of some metal deposited on the silvering. A convenient space on the bulb is left clear to allow light to enter, and down the middle of the bulb a rod is fixed. Finally the rod and the thin metal lining are connected so that a continuous " circuit " is provided for the flow of electricity round the cell. As light enters the tube, electrons are thrown out from the metal, and some of them travel across to the central rod and so back, *via* the external connection, to the metal film from which they started. A photo-electric cell is in fact nothing more than a convenient way of harnessing the throwing out of electrons by light so as to produce a continuous electric current so long as light is falling on the cell.

Such in outline is the photo-electric cell which, with its current suitably amplified, is capable of an almost infinite variety of robot achievements. It can be made, for example, to switch on street lamps when any chosen degree of darkness is reached, to sort packages according to their brightness, and to operate a burglar alarm when the entry of a burglar cuts an invisible infra-red ray. But what we are here concerned with is that by providing the necessary translation of light into electricity it makes television possible.

Given the photo-electric cell as a tool, it may make for clearness if we now put down in order the three things which the television engineer must accomplish if he is to produce practical television. We can then see how far recent developments have enabled him to do them more effectively. He must—

(1) Devise a satisfactory method of dividing up a picture into units, the light and shade of which can be *individually* translated into electric currents. This is essentially the same problem as that of putting the picture together at the receiving end, although in practice it does not necessarily follow that the same method will be used in both cases.

(2) If television is to be broadcast, he must be able to put the necessary radio signals across. In doing this he must so far as possible keep clear of all atmospherics and other sources of outside interference, because stray extraneous noises are very much less noticeable

to the ear than when translated into miscellaneous "splodges" sprawling across the face of a picture.

(3) Manage to keep transmitter and receiver moving in exact step. In other words the picture on the receiving screen must be synchronised with the transmitter if it is to be kept steady.

The practical solution of the first and in many ways the most important of these problems was provided in 1926 by Mr J. L. Baird, the inventor son of a Scottish minister. It was Mr Baird who first achieved true television as opposed to the transmission of outlines; and, whatever the outcome of the present struggle for television supremacy, posterity will have to concede to him a similar position as practical pioneer to that occupied by Marchese Marconi in the story of the radio wave.

It would, of course, have been theoretically possible to arrange that each section of the picture was separately "viewed" by an individual photo-electric cell and the electrical record of each section's light or shade independently transmitted to the receiving end. But a moment's thought will show that any arrangement of that kind would be impossibly complicated and expensive, and that an almost infinite range of wireless wave-lengths would be necessary to transmit a picture in anything approaching fine detail.

Practical television, like cinematography, depends on the fact that the eye is a deceiver. In a film the eye is hoodwinked into mistaking a succession of separate pictures for continuous motion. In television the same process of deceit is carried a stage farther. Not only are successive pictures separate and stationary, but the eye is only shown one small part of each picture at a time. Yet different sections of the picture arrive in such rapid succession that the eye has not forgotten the first section of the picture before the last section has arrived. It sees the picture as a whole. From another point of view television may be described as a cross between picture reproduction as achieved in film and newspaper practice. Film reproduction supplies the idea of a succession of pictures; newspaper reproduction the idea of dividing a picture up into a large number of small areas, in this case dots, within which all fine detail is ignored. As in a film, the stream of pictures must be fast enough to avoid flicker, and as in a newspaper the grain of the pictures must be fine enough to carry conviction.

Mr Baird's great contribution to television was his success in "scanning" with a moving spot of light the picture or scene to be transmitted. The essence of this arrangement is that, while the whole picture is within range of one or more photo-electric cells, these cells are only stimulated at any given moment by the light from one small area. This is achieved by arranging that while the general lighting is relatively faint, the illumination provided by the spot of light is very bright; so that a single group of cells connected to a single radio channel can be made, in theory at any rate, to transmit the whole of a complicated picture in as great detail as we care to arrange.

It is also, of course, necessary to arrange that the spot of light travels over the picture in a regular way, duplicating nothing and missing nothing. In practice the spot is made to traverse the picture in a series of vertical strips, starting at one edge and always moving downwards along successive strips until the whole picture is covered. This has been done in two different ways, the two pieces of apparatus being known as the "scanning disc" and the "mirror drum"; but as both these devices merely represent ingenious mechanical arrangements for securing the desired motion of the light spot I do not propose to describe them here.[1]

We have now put our photo-electric cells in a position to televise. At any one moment they are only going to be illuminated by the light reflected from one section of the picture. But provided the picture is scanned sufficiently rapidly they will be able to translate the varying light and shade of each strip of the picture into a continuously varying electric current before the eye has had time to notice what is happening. This rapidly varying current, which bears in itself the whole imprint of the picture, is then passed out into space as a radio signal of similarly varying intensity. That is the half-way stage between picture and receiver, and the second set of metamorphoses leading from radio wave to televised picture is now begun. First the movements of the incoming radio wave are caught and amplified by an ordinary wireless receiver. The result is an electric current which exactly corresponds with that originally produced by the photo-electric cells. Next the neon lamp, already referred to, is brought into action to translate these electrical changes into light changes. And finally, by a mechanical arrangement identical with the "scanner" at the

[1] Both these arrangements had, incidentally, been suggested as possible mechanisms for use in television nearly thirty years earlier.

transmitting end, the changing light from the lamp is made to traverse the screen of the home televisor. As the brightness of the neon lamp varies, each area of light or shade is allocated to its correct position on the screen until the whole picture has been built up.

That, in broad outline, is the scheme of the experimental television broadcasts undertaken by the B.B.C. in 1932. It has been necessary to describe it in order that we may see what improvements modern developments make possible. In the meantime we may notice that in this system the pictures are only divided into 30 strips which is not enough to give good definition, except in extreme "close-ups"; and that only twelve or so successive pictures are provided a minute, which is not enough to give a feeling of continuity. The reason for these shortcomings is that, on ordinary broadcasting wave-lengths, there is simply not enough room to transmit pictures in finer detail. It is a question of the same "side band" spread which we met in discussing the amount of space available at different wave-lengths for the transmission of sound. Only the difficulty, in the case of television, is very much greater.

The necessary condition for the production of "side band" spread, it will be remembered, is that an extra message-carrying vibration should be imposed on the natural frequency of the radio waves. In television the imposed vibrations correspond with the light and shade of the picture as it is scanned by the moving spot of light. Without going into detail it is clear that the more finely we scan the picture, and the more pictures we send over a second, the more rapid will these imposed vibrations have to be. So that the higher the standard of reproduction which we set ourselves, the more space must we be allowed in the ether in which to put our pictures across.

In practice the amount of radio space required also depends on the amount of detail in the pictures to be transmitted. At one end of the scale a plain sheet of white paper would provide no contrast at all; and at the other a finely drawn checkboard, with squares corresponding exactly in size with the moving spot of light, would provide as much contrast as that particular televisor could handle. The radio space necessary if such a checkboard were to be sent over could be precisely calculated. But it would be a futile pastime because no real life picture would ever be so exacting. It is simpler to accept as an experimental result the fact that the original 30-line transmissions in this country just about fitted comfortably into the space available round the London Regional transmitter. It represents the limit then set to television progress by the requirements of broadcasting programmes.

This is why the whole position has been revolutionised by the advent of ultra-short waves. There is more room for television transmissions between 6 and 6·2 metres than there is within the whole range from 150 to 600 metres; and this, as it happens, would be roughly the space required to send over the latest pictures, with 180 line scanning instead of 30 line, and 25 pictures a second instead of 12. Such transmissions give nearly forty times better definition than was obtainable with the old system, while the greater frequency of the pictures greatly increases the feeling of steadiness. Two hundred and forty line scanning, which is already being tried, gives a standard of definition which is probably rather better than that of the home cinema film.

The reception of ultra-short waves also satisfies the condition of being absolutely free from atmospherics, while a recent demonstration in London has shown that the anticipated bugbear of motor-car interference is adequately removed, even in a busy street, by placing the receiving aerial on the roof.[1] There remains the question of the range of reception obtainable, because it will be remembered that ultra-short waves travel in practically straight lines and are too short to make use of the radio mirrors. In the case of an experimental transmitter, mounted on the south tower of the Crystal Palace which has the advantage of standing on high ground, the range has been explored in a series of tests made with a lorry carrying a portable receiver. The results showed that full-strength reception could be obtained as far north as Hatfield and as far south as the heights of Ashdown Forest. A receiver in the latter area would naturally be exceptionally well placed, but it looks as if an area bounded, at any rate, by Hitchin on the north (37 miles), Ilford on the east (13 miles), Staines on the west (19 miles), and Cobham and Caterham on the south (16 and 10 miles respectively),

[1] This was satisfactorily proved at the first public demonstration of television using ultra-short waves and cathode-ray-tube reception. The demonstration was given by the Baird Company on March 20, 1934, transmission being from the Crystal Palace and reception in Wardour Street where traffic is notably heavy. It is largely due to the courtesy of the Baird Company that I have been able in the following pages to give so complete a picture of what, at the time of writing, are the latest developments in this sphere.

could be well served from the Crystal Palace. The needs of Greater London could therefore be met by a single transmitter, while it is estimated that half a dozen or so transmitters placed in heavily populated areas could reach 75 per cent. of the inhabitants of England, Scotland and Wales.

As compared with ultra-short waves, micro waves would only have the advantage of being completely immune from motor cars and other sources of electrical interference. Since they have never yet been generated in anything approaching sufficient power for broadcasting it is probable that they will make their contribution to television in another way. Micro waves may be expected to replace the land-lines which the B.B.C. use, in the case of sound broadcasts, to connect studios and transmitters and to link up their different stations. Land-lines are unable to handle, except over very short distances, the rapidly changing electric currents needed to convey picture images in fine detail. But micro waves have already been used in this way at the Crystal Palace, and, within a few years' time, it looks as if their use will be very much extended. For example, if the Derby was being televised at Epsom, it would be possible to flash the pictures through a series of micro-wave relays to the regular ultra-short wave stations in London, Manchester and other centres; and as micro-wave transmitters can be made relatively small, the necessary relay stations could be carried about the country on lorries which would be temporarily parked in whatever positions proved best for the job in hand.

We now come to the second great impetus which

modern research has given to television. It has been provided by the instrument known as the cathode-ray tube, which we have already seen has proved so useful in charting the positions of the radio mirrors. The cathode-ray tube, it will be remembered, produces a moving beam of electrons which, falling on a fluorescent screen at the end of the tube, appears as a spot of light. The phrase "spot of light" at once suggests its usefulness in television; and in fact it provides what is probably the best solution to the last of the great problems of the television engineer, that of keeping transmitter and receiver in step.

It has been mentioned already that the moving spot of light used at both the transmitting and receiving ends in the systems of television previously described is controlled by mechanical means. There are moving parts which revolve at high speeds and, if the picture is to be kept steady in the receiving screen, the speed of movement at transmitter and receiver must somehow be kept identical. The problem is to keep a series of small electric motors in the home of every receiver moving in exact step with a similar motor at the transmitter. Obviously this is not an easy thing to do and, however much progress has been made in the last few years, it is impossible that it should ever be done perfectly. The transmitting motor is bound, at times, to vary a little from its scheduled speed as the current supplied from the mains alters, and even if the receiving motors are told by wireless of what is happening they cannot respond at once. They have weight, and anything which has weight, for example a motor car, can neither be checked nor speeded up instantaneously.

The beauty of the cathode-ray tube is that the electrons which provide the spot of light have no weight, or at any rate no appreciable weight. They are the lightest particles known to science. So if the transmitting motor is being a little erratic they will respond to this erratic behaviour as soon as they are told to do so. Moreover, used as a receiver, the cathode-ray tube replaces lamp, scanning disc and screen at one swoop. The strength of the electron beam varies in response to the radio signals just as does the brightness of the neon lamp; the path traced out by the moving beam can be made to follow exactly that of the scanning spot at the transmitter; and the fluorescent screen at the end of the tube serves to show the picture directly. The greatest size available at the time of writing is 10 inches by 8 inches, but this can be magnified by means of a lens and there seems no reason, if so desired, why the picture should not ultimately be projected on to the type of screen normally used in home cinematography and perhaps even on to the still larger screens of film palaces.

In projection on to a screen, as in all high definition television, one of the essential problems is to produce a bright enough picture. We meet this difficulty at both the transmitting and receiving ends, and it arises from the fact that with 180-line scanning the size of the moving light spot is less than one forty-thousandth of the whole picture. This means that each individual section of the picture is only being transmitted for one forty-thousandth part of every second, although the eye has got to see the whole of the picture the whole of the time. Actually the light used in the scanning process has to be five hundred times as strong as was sufficient for the old thirty-line system, and even so it has not yet proved possible to show distant "shots". At the receiving end, matters are helped somewhat by the fact that the fluorescence of the cathode-ray tube continues for quite an appreciable time after the electron beam has passed on to different parts of the screen. The inherent brightness of the screen has also been greatly increased in recent years, and it may be expected that in time screens of still greater brightness will be produced.

At the best, however, it is probable that a large proportion of television transmissions will be made with a cinema film as an intermediary, and at least one would-be commercial system is definitely confined to use in this way. It is difficult to argue that it is any serious disadvantage for a home audience to have a cinema film interposed between a real life scene and what they see on their receiving screens. The delay involved is negligible. The whole process of photography, developing and transmission can be worked continuously and takes a matter of 10 seconds from start to finish. As sound films are used, sound and vision are as perfectly synchronised as in any "talkie", and the only difference would be that the audience would see the finish of, say, the Boat Race about a sixth of a minute after it really took place. In any case film television has every advantage for outdoor events, in view of the very special difficulties which would in most cases be involved in securing sufficiently powerful lighting for direct transmission.

Naturally there is even less difficulty in transmitting

an ordinary cinema film, although at the present stage of development it looks as if a special technique would be involved in the making of suitable films. At any rate, judging from demonstrations which I have seen, it is clear that all films are not equally suitable. Close-ups are almost perfect, and middle distance scenes come out quite effectively. But distant action and dependence on background effects would be best avoided at the present stage of development. It also seems probable that more emphasis could usefully be placed on the spoken word than is the case in the average talkie. Such temporary limitations are in any case only due to the relatively small size of present receiving screens. As cathode-ray tubes are made larger and brighter, and projection on to an outside screen becomes possible, there is no doubt that the cinema industry will find itself confronted with a serious rival. Germany has indeed already produced a television system in which a cinema film is interposed at the receiving, as well as at the transmitting end.[1] In this case the second film is, literally, "shot" by radio and, if transmission were good enough, would presumably be indistinguishable from the genuine article. Whether the two industries will operate as friends or enemies only time can show. From the point of view of production costs there can be no doubt that alliance would be preferable. Nor is it reasonable to suppose that the cinema will ever be killed by tele-

[1] The total delay involved in the developing, fixing, washing and partial drying of the two films is less than three minutes. In the interests of economy the film at the receiving end is used again and again, being continuously washed clear and re-emulsified between one set of exposures and the next.

vision. The cinema hall will always have its attractions. "Going out" is more of an event than staying at home; and mass emotion is, for many, a better relaxation than fireside criticism. There is no logical reason why both forms of entertainment should not flourish together.

As might be expected in so youthful an industry there are also in the field a number of other systems which are entirely different from any of those already described. But none has yet reached the stage of practical development, and there is little doubt that the immediate future lies with a combination of mechanical "scanning", ultra-short-wave transmission and cathode-ray-tube reception.[1] There is, however, one other embryo television system which is of very special interest. Its great merit, if it proves practical, will be the immediate solution of all illumination problems. Indeed it seems to represent the only hope of ever securing direct and satisfactory television of real-life scenes under natural lighting conditions. All this arises from the fact that the whole of the picture to be transmitted is being recorded the whole time.

This truly remarkable system comes from America. It is the invention of Dr V. K. Zworykin, who has devised a light-sensitive screen which is entirely different from any previous arrangement. The general idea is that a small thin sheet of non-conducting material is

[1] There is some theoretical advantage to be gained from using a cathode-ray tube for "scanning" at the transmitting end as well. But it does not appear that satisfactory performance under working conditions has yet been achieved. Such a system is essentially limited to film transmissions.

The chief drawbacks to the cathode-ray tube are the expense of replacement and the fact that it is still far from being fool-proof.

covered with an enormous number of fine metallic particles, each one of which acts as a small photo-electric cell on its own. For the time being there is no flow of electric current, but at each point of the mosaic an electrical charge is continuously being built up which accurately corresponds with the varying light and shade of any picture which is focussed upon it.

Scanning is performed by the moving beam of a cathode-ray tube within the glass walls of which the light-sensitive screen is placed. The function of the cathode-ray beam is to release the accumulated electric charges, which have been built up at each point of the mosaic since the scanning beam last passed that way. The resulting electrical impulse is then used to modulate the signals sent out by the radio transmitter in the usual way. The theoretical benefit obtained by the continuous recording of the whole picture the whole time exactly corresponds with the number of units into which the picture is divided for transmission. With the system, as at present used by Dr Zworykin, the brightness should be increased 70,000-fold if his minute photo-electric cells were as efficient as the standard variety. He claims to have realised about a tenth of this benefit under working conditions, quite a big enough gain to be of revolutionary importance.

If Dr Zworykin's invention is successfully developed, it will mean that it will be possible to televise any outdoor or other scene which is sufficiently brightly illuminated to be cinematographically recorded. But it must be remembered that the whole technique is entirely new and that there must inevitably be a great many practical problems to be solved before satisfactory reproduction can be secured. Television of the type already described is certain to be first in the field and it will remain to be seen how much the advantage of direct transmission, without the use of a cinema film record as an intermediary, will appeal to the popular imagination.

Again, both coloured and stereoscopic television, in which the picture would be seen in relief, are as much practical possibilities as similar developments in the cinema world. A colour television process of a kind was indeed produced as far back as 1928, but the whole trend at the moment is, very properly, to concentrate on producing really satisfactory "straight" television without worrying unduly about further complications. It may be expected that the pace in this matter will be set by the film industry. The advent of colour films must, sooner or later, mean the advent of colour television.

CHAPTER VIII

RADIO AND MEDICINE

W<small>HEN</small> ultra-short waves were first discussed, it was mentioned that their properties more nearly approached heat waves than those of ordinary wireless. It should not therefore be surprising to learn that it is possible to cook an egg by wireless. It may be more of a surprise that ultra-short waves can cook the yolk and leave the white untouched.

Behind this apparent miracle lie all the medical possibilities of ultra-short waves. Let us think for a moment of what it implies. It means that the waves in question have passed through the white without being appreciably absorbed, only to yield up a considerable fraction of their energy when they reach another kind of matter—enough, at any rate, to produce a perfectly good hard-boiled yolk.

It only remained for scientists to discover that different wave-lengths of ultra-short wireless were absorbed by different kinds of matter for their possible usefulness in medicine to be at once evident. For example, if a wave-length could be found which would be absorbed by the lungs and not, say, by the heart, it should be possible to affect the one organ without affecting the other. It does not necessarily follow that the effect should be a good one. But here at least is something which should be worth investigating. As a first picture, and keeping to the example of the egg, we might expect that it would be possible to apply something not unlike an internal hot poultice to selected parts of the body.

There is nothing in the least impossible in the idea that different kinds of tissue might absorb particular wave-lengths. In physical science it has been known a long time that every kind of matter, according to its momentary condition, will only either absorb or emit its own special wave-length. Some atoms, for example those of gases, are excessively pernickety, and insist on exactly specified wave-lengths. With matter in other conditions it may merely be that radiation within a particular region of wave-length is more readily absorbed or emitted than any other wave-length. The first kind of idiosyncrasy tells astronomers what the stars are made of and whether they are moving towards us or away from us, while the second kind of idiosyncrasy enables their temperatures to be calculated. Astronomers, as might be expected, deal mainly with light, but the same peculiarities can be traced for every kind of radiation from the shortest waves of X-ray to the longest waves of heat.

This liking for particular wave-lengths has even been lately detected in the human body. There is an exact parallel in the way in which nature provides man and animals with vitamin A. This vitamin, which is sometimes called the anti-infective vitamin, is made by both chemists and nature from a material called carotene, which derives its name from being the colour-matter of carrots. Evidence has been obtained both at Liverpool and at Cambridge, that a particular wave-length is necessary to bring about the change from carotene to vitamin. Although it has not been directly proved, it is fairly certain that this particular wave-length, obtained from sunlight, is what enables an animal to make vitamin A. So whether man makes his own vitamin A, or takes it ready-made in cow's milk or butter, he is dependent directly or indirectly on a particular wave-length in the sun's radiation. That, incidentally, is why milk is less rich in vitamins in winter than in summer. Less sunlight is available for their manufacture.

This is the sort of way in which the problem of medical radio first struck Dr Erwin Schliephake, the young German doctor who ten years ago began the first experiments in this field. The ten years period is important. For three years Dr Schliephake, *privat-dozent* of Giessen and Jena Universities, experimented only on animals. For seven years more he has worked on human beings. It cannot be said, therefore, that his claims are in any sense based on hurried work. In fact they are both modest and well-substantiated, and to a large extent have been confirmed by others.

At first sight the heating effect obtained by ultra-short waves might seem to resemble closely that obtained in diathermy—the process of heating the body of the patient by passing ordinary high frequency currents through it. There are, however, some very important differences. Unfortunately for diathermy the fatty layers which surround the body and keep it warm are relatively resistant to the passage of ordinary electric currents. By far the greater part of the heating effect is therefore used up in the outer layers of the body, while only a small amount of the total heat produced penetrates within. Here ultra-short waves are at once in marked contrast with diathermy. The extent of heating depends solely on the capacity of particular tissues to absorb them, and it so happens that only a relatively small amount of the total energy is used in penetrating through the outer layers. It is therefore possible to heat the inside of the body more than the outside, just as the yolk of the egg was heated more than the white.

Even inside the body, the advantage is still with the ultra-short waves. The currents of diathermy, like any other electric currents, tend to follow the path of least resistance, that is to say chiefly along the blood vessels. Dr Schliephake has, however, been able to prove both by laboratory experiments and on human beings that the membranes of the body are also penetrated. In addition the heating effect is actually greatest where the circulation is least, and that for quite a different reason.

The blood stream, like the circulating water of a car's radiator, is always doing its best to keep the body temperature uniform and even, a task which it performs with considerably more efficiency than does the best of radiators. The result is that all the time that ultra-short waves are applying their internal hot poultice the blood stream is carrying away part of the heat produced. Consequently, wherever, as in a tumour or ulcer, the normal circulation is interfered with, the heat produced has a chance to accumulate. Dr Schliephake has also been able to prove that the smallest particles are directly reached—for example, that the blood corpuscles are heated to a greater extent than the surrounding serum.

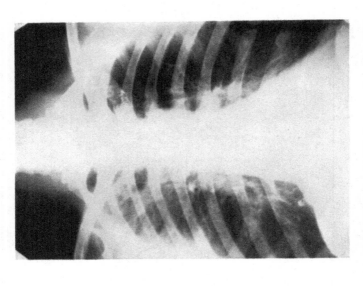

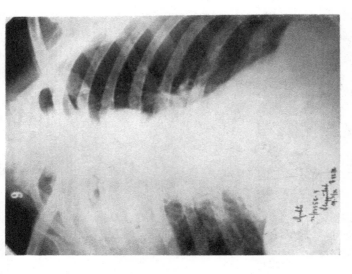

Radiograms showing the clearing up of purulent matter in the neighbourhood of the lungs under ultra-short wave treatment—24 days interval

The use of ultra-short waves to cure disease began modestly when Dr Schliephake got the opportunity to treat a furuncle (type of boil) on his own neck. It was a modest beginning because the chief virtue of ultra-short waves is their power to penetrate well within the body. None the less, it was sufficient to prove that they could be safely and successfully used, and Dr Schliephake has since treated hundreds of similar infections in many other parts of the body in the same way.

Some of Dr Schliephake's most impressive results have been obtained with lung abscesses. He was recently able to report the cure of twenty cases of lung abscess and related complaints. They were all cured without operation. What this means in terms of human life can fortunately be rendered in cold figures. Without operation the usual mortality is between 65 and 75 per cent. With operation, the mortality is about 40 per cent.; and even then a certain proportion of the patients who have been operated upon may remain chronic invalids. The photographs reproduced illustrate another type of chest treatment. They are a clear indication, even to a lay observer unused to the examination of X-ray records, of what ultra-short-wave therapy can do.

Pyorrhoea, the disease of the gums to which we are so often reminded that four out of five people fall victims, is another field in which this form of treatment has been successfully used. So are rheumatism and various forms of arthritis. But in the case of rheumatism it has proved important that the focal centre from which poisoning proceeds should be diagnosed and directly treated. For example, if the trouble originates in the teeth, then the dental abscesses have in some cases been cleared up and the patient spared the premature instalment of false teeth.

It is likely that ultra-short waves will prove equally important as a non-operative treatment for certain cases of duodenal ulcer. Encouraging results have also been obtained with asthma, migraine (headaches of unknown origin), osteomyelitis (which causes so many cases of chronic crippling), and purulent inflammation of the various nasal sinuses (small air cavities connected with the nose).[1] Nor does this short list by any means exhaust the possibilities.

Finally there is an entirely distinct application which depends on the power of intense radiation to heat the blood stream generally and so produce an artificial fever. This is a form of treatment used in dealing with that most terrible scourge, general paralysis of the insane. Artificially induced bouts of malaria have been used for some years in the treatment of this disease, the general effect being very similar. At first sight it might seem that the balance of advantage was entirely with the new treatment. The rise in temperature, being externally produced, can be exactly controlled. On the other hand a higher powered apparatus is required which is not only expensive to manufacture and in use, but introduces the possibility of surface burns. The treatment is also exceptionally exhausting to the patient.

For these reasons Dr Schliephake prefers a com-

[1] Dr Schliephake instances empyemae of the antra, frontal and even the ethmoidal sinuses.

promise of his own. The high power needed for the "general fever" type of treatment cannot be produced at present on wave-lengths of less than 30 metres. Dr Schliephake uses instead a rather less powerful 12-metre transmitter with which he can combine local treatment of the head with a milder general fever— a rise in temperature of only two or three degrees compared with the five degrees rise which can be produced using longer wave-lengths.

In the last resort it is the wave-length used that seems to be of primary importance. Selective effects, Dr Schliephake reports, are only obtained within the range of roughly 16 to 1 metres. Without claiming any finality for his figures he states that he has found a wave-length of 4–6 metres most effective in reaching the nasal sinuses, while with asthma and migraine good results have also been obtained with waves of from 10 to 15 metres. These are merely quoted as examples to show that there do appear to be specific differences between the effects of different wave-lengths.

Enough has, at any rate, been written to show that ultra-short waves present a most important field for medical inquiry. At the same time the technique is probably at least as complex as that of X-rays, although for the patient the only sensation is one of pleasant warmth. For this reason it is not to be expected that sensational results may be at once expected in this country.

At least three London hospitals are now engaged in ultra-short-wave experiments, and no doubt the near future will see the usual succession of claims and denials which follows the introduction of almost any new development in medicine. As in any form of treatment which involves complex apparatus, the doctor is more important than the method. In the case of ultra-short waves the factors which can be varied include not only the wave-length, but the intensity of the waves, the length and frequency of treatment and—most important of all—the adjustment of the "electrodes" which to a large extent determine the path which the waves take through the patient's body. It is not to be expected that miracles should be worked without a long apprenticeship.

All that can be said at the moment is that ultra-short waves may be expected to have a place in the medical armoury at least as important as that of X-rays, and probably more important. A great deal depends on the extent to which it proves possible to elaborate the specific characteristics of individual wave-lengths.

Dr Schliephake's belief that the direct effect of irradiation is probably on the bacteria which cause disease is, in many ways, hopeful. Certainly there is a wide field here for investigation, even if it be admitted, as it must be, that the power to kill or weaken bacteria in a test-tube is by no means equivalent to the power to do so in the human body. This is the real gulf between science and medicine and the real and most charitable explanation of medical conservatism.

Finally, a word on the all-important specific effect. There is reason to believe that this is due, not to the ultra-short waves themselves, but to the rapidly changing electrical forces which accompany their production. This does not seem to be a distinction of more than strictly scientific interest. For practical purposes it is enough that the use of a special type of wireless transmitter has produced medical results which have never been realised in any other way. Equally, on the theoretical side, the most significant fact is that particular tissues of the body do appear to respond in a special way to electrical changes of particular frequencies.

CHAPTER IX

RADIO AND SAFETY AT SEA

Nowhere have the benefits of wireless been so practical and unquestioned as on the sea. There are those who still protest that they would prefer the peace of a pre-broadcasting evening to any programme which the B.B.C. is ever likely to give them. There can be no one who can honestly say that he would prefer, when at sea, to be without the seafarer's final refuge—a broadcast S.O.S. The first demonstration of the power of radio to save life at sea was given as long ago as 1899 when a steamer collided with the East Goodwin lightship. The lightship, which was the first to carry a radio transmitter, was able to report the accident to the South Foreland, and lifeboats were at once sent out. How many lives have since been saved by similar appeals for help it is impossible even to estimate, although the necessary calculations will no doubt one day be made by some enterprising statistician.

There are, of course, many other advantages available to the great body of ships equipped with wireless, although it may be doubted if the change which has been wrought is fully appreciated by the average passenger. One of them, the printing of an up-to-date ship's newspaper, dates back almost as far as the East Goodwin S.O.S. It was in 1900, during the South African War, when Marconi was travelling back from America to Britain, that he was asked to make an attempt to get the latest war news before the liner *St Paul* docked at Southampton. He was able to get into communication with a radio station at the Needles while the liner was still in the English Channel and the first "Atlantic" newspaper, admittedly of rather meagre proportions, was rushed out with a few hours to spare. Nowadays ship-to-shore telephony is taken almost for granted by those who can afford it, and only a faint stir of interest is aroused when a passenger casually learns that the doctor of his ship has been giving medical advice by radio. It quite often happens that illness on a smaller boat is diagnosed in this way and suitable treatment prescribed; and, as a last resort, arrangements can be made for the liner to divert its course and take the patient on board for an operation. Most of all, perhaps, is it difficult to realise that, little more than thirty years ago, every ship at sea was irrevocably cut off from the outside world; and that, in the case of our own island, it was quite usual for incoming freighters to call at the Scilly Isles to learn at what port their cargo was to be unloaded.

The silence of the seven seas has been broken, and radio is now well on its way to overcoming what is still their greatest danger, that of navigation in fog. The remedy, as might be expected, lies in recording the direction of incoming radio signals, both from beacon stations on shore and from other ships. It is a remedy which has not yet been fully applied, but the scope for its application is almost unlimited.

The simplest lesson in radio direction-finding can be learnt at home with any ordinary portable receiving set. It is a familiar fact that as a radio receiver of this kind is turned round the strength of reproduction changes. This is because the aerial in a portable set is in the form of a closed loop which will only receive radio signals efficiently if the loop is in a straight line with the transmitting station. So, if one happened not to know in what part of Europe Huizen or any other station was, it would be possible to obtain a rough idea of, at any rate, its direction, by simply turning a portable set round until reception was best or worst.[1] The first position would mean that the aerial loop was pointing at the wanted station. The second position, which can be measured more accurately, that the aerial was at right angles to the direction of the station.

Every form of direction-finder at present in practical use is based on this simple principle. In every case either the aerial system is rotated, and the position of minimum signal strength directly noted; or, if the aerial system is kept stationary, then some other part of the apparatus is rotated which produces a corresponding effect. The direction from which any kind of broadcast signals are coming can be equally well diagnosed, and all that is necessary is that a special type of receiver should be used in order that the record obtained should be as accurate as possible.

Even the need for special receivers can, however, be eliminated if a directional aerial is used at the transmitting end. This is the principle on which the "rotating loop" type of radio beacon works. Just as a loop aerial, when used as a receiver, is least sensitive

[1] See, however, p. 160. The difficulty, there described, might prevent a correct indication being obtained of the direction of the transmitting station if the test was made during the hours of darkness.

to waves approaching it at right angles, so when a similar aerial is used for transmitting, there is a "band of silence" stretching out on either side of it. In practice it is arranged that the transmitting aerial makes one complete revolution a minute, and that the direction of minimum strength is as clearly marked as possible. At the moment when signal strength is weakest in a north-south direction, a special signal is sent out. So all that any listening ship need do is to

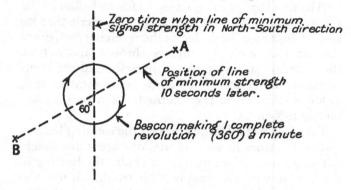

Fig. 19

time the interval between a north-south signal and the moment when reception is weakest. An interval of 10 seconds, for example, would mean that the transmitter had made one-sixth of a complete revolution (i.e. turned through 60 degrees) in a clockwise direction from the north-south position. Fig. 19 will make this clear.

The only possible complication is that the navigator has no direct way of telling in which of two opposite directions the beacon lies. The ambiguity, which is inevitable in any form of radio direction-finding,

arises in this case from the fact that the time of minimum signal strength stretches out equally on either
side of the rotating aerial. But where the transmitter
is on shore and in a position known to the navigator it
is hardly likely that he will have any doubt about
which is its true direction.

There are two radio beacons of this kind operating
on the English coast, one at Orfordness in Suffolk and
the other at Tangmere near Selsey Bill, and it may
well be that the system has a useful future before it. Its
range is about 200 miles and its great merit that no
special apparatus is needed at the receiving end, other
than an accurate stop-watch or chronograph to time
the moment of lowest signal strength. All the complications are concentrated at the beacon station and,
so far as the receiver is concerned, the system is completely fool-proof.

Yet in spite of all its advantages the rotating beacon
system has from its very nature, one great drawback.
Although it enables any ship to obtain the bearing of
the beacon with the least possible trouble, it tells the
navigator nothing whatever about the position of other
ships. In other words it will prevent a ship getting lost
in fog, but it will not prevent it colliding with another
ship. If the latter and, in many cases, more important
type of security is to be achieved all ships must be prepared both to carry specifically direction-finding equipment and to send out regular radio signals during fog.

Still the main problems of direction-finding look
simple enough at first sight, and a natural question to
ask is why radio navigation has not been universally
adopted. The answer is to a large extent supplied by

Fig. 20

Map showing distribution of radio beacons around the British coasts.

the last eccentricity of the radio mirrors in the atmosphere which we shall have occasion to notice.

Both these mirrors have an unfortunate habit of imparting a "spin" to any waves which they reflect. Without going into details it is not difficult to see that the effect of this spin may be to give a false impression of the direction from which the waves are coming—particularly if the reflected wave is being strongly received in comparison with the direct wave, or if the reflected wave is coming down at a steep angle.[1] The effects on direction-finding can be very serious, particularly over short distances. A known stationary transmitter at Teddington, for example, has been observed to make six complete revolutions, to all appearances, in the course of one evening about a receiver at Slough ten miles away. Although admittedly an extreme case, this is the sort of performance which modern direction-finding has had to live down in order to establish its reliability.

One way of getting rid of the unwanted effects of this "wave spin" is to use a more complicated type of aerial than the plain rotating loop which constitutes the simplest type of direction-finder. The general idea

[1] The way in which the correctness of this explanation has been verified is not without interest. It has been found that the spin of the reflected waves is predominantly left-handed in the northern hemisphere and predominantly right-handed in the southern. Whether transmissions are from north to south, or south to north, makes no difference. The contrast depends on the fact that the earth's total magnetic "field" is always downwards in the northern hemisphere and upwards in the southern.

The original records of contrasting spin were made in England by Professor Appleton and the Cambridge scientist, J. A. Ratcliffe; and in Australia by A. L. Green, working under the Australian Radio Research Board.

is to make the twist of the reflected waves work in one direction in one part of the aerial, and in the opposite direction in another part. Then if the aerial is appropriately connected to the receiver, the net result will be two equal and opposite irregularities which cancel out. It is an immoral system, but it works. That is the way in which any form of direction-finder of the type already considered can be made accurate. It is also the only possible system which can be applied to ordinary radio signals not specially designed for direction-finding purposes.

There is, however, one other possible way of eliminating "wave spin" irregularities which is of very special interest because it leads immediately to what looks to be an eminently practical method of preventing collisions in fog. In this case the idea is to ignore all reflected waves completely, and only make use of those waves which have pursued a direct course from transmitter to receiver.

One of the most useful ways of charting changes in the radio mirrors, it may be remembered, depends on just such a separation of the direct and reflected waves. It was achieved by the now inevitable cathode-ray tube and involved a special transmitter, sending out short "pulses" of energy lasting only about one ten-thousandth of a second. The point of using these short pulses was that the disturbance of the cathode-ray beam caused by the arrival of the direct wave was already over before the reflected wave, making a slightly longer journey, had had time to arrive. The two sets of waves cannot be separated if the signals radiated are continuous, but can be separated if the signals are sufficiently short.

The application of this principle to inter-ship direction-finding is at once obvious. While it is easy enough to say that every ship at sea during fog ought to send out radio signals to be "direction found" by its neighbours, the practical difficulties would, in the ordinary course, be insuperable. Short of allotting a separate wave-length to every ship it would be impossible to avoid mutual interference. But if the signals sent out lasted only one ten-thousandth of a second at intervals of, say, ten seconds the chance of their being simultaneously given from two neighbouring ships would be infinitesimally small. Every ship could therefore send out its position-bearing signals on the same wave-length at the same time without fear of causing confusion; and they could even use their transmitters simultaneously for Morse signalling since the direction-finding "pulses" would be too short to cause interference with any messages imposed upon them.

Here, then, is the perfect solution of ship-to-ship direction-finding so far as the transmitting end of the problem is concerned. We must now see how the cathode-ray tube solves the receiving problem as well. The direction-finding application depends once again on the capacity of the electron beam for being pulled about. As usual we start with the beam concentrated into a single spot of light on the fluorescent screen at the end of the tube. All that is necessary is to arrange that the signals received from one aerial pull the spot out in one direction, while those from the other aerial pull it out at right angles. The ground-wave signals,

*Radio direction-finding signal as shown by
a cathode-ray tube*

which arrive first and are free from any appreciable errors, give a long straight line which accurately represents the direction of the transmitter. Then, when the dancing electrons have finished tracing out this straight line on the fluorescent screen, the reflected waves begin to arrive. This time the pattern produced is more irregular, generally an oval-shaped "splodge" in the middle of the straight line. The eye, of course, sees both patterns at the same time—the wanted straight line and the unwanted irregular patch. But as the straight line extends far beyond the limits of the central patch there is not the smallest difficulty in reading the true direction from it. The photograph here reproduced shows how clear the indication of direction is; and, which is of equal interest for small ships, the necessary receiver is of convenient size. The aerials need only be about 4 feet high, and the whole of the rest of the apparatus could be comfortably enclosed in a good-sized suit-case. Even in a coastal steamer it should be possible to store this most compact direction-finder conveniently on deck.

What, then, does the instrument tell the navigator? In the first place successive flashes show the direction of every ship within a five or ten miles radius[1] according to the strength of their wireless transmitters, although from a single signal it is not possible to distinguish between pairs of opposite directions. Doubt on this admittedly important point is removed with almost complete certainty at the second appearance of the

[1] In the unlikely event of signals from two ships arriving at the same instant they are still readily distinguishable.

flashing green line. If the line is longer than at first, the signal received must be stronger—that is, the two ships are nearer. This, in itself, is information of the utmost value and in most cases will enable the navigator to tell in which of the two possible directions the other ship is.

If, as well as increasing in length, the tell-tale line remains always pointing in the same direction, it follows that the two ships are not only getting nearer, but are definitely heading for a collision. This is probably the greatest security which this form of direction-finder gives. So long as the direction of the line changes by a reasonable amount the navigator can feel secure.

Even now, no reference has been made to what would in practice be the greatest safeguard of all. Since both ships are necessarily equipped with ordinary Morse signalling, their respective captains would naturally communicate changes of course to each other. The results as shown by the direction-finders would remove any lingering doubts as to where the other was, and would also prevent any possibility of that distressing form of last-minute dodging when two people meet at a street corner.[1]

Exactly the same system could be applied to transmissions from radio beacons, of which there are already

---

[1] Before leaving this application of direction-finding we may notice that infra-red rays hold out the possibility of yet another form of collision-preventing device. It looks as if it should be possible for ships to obtain a continuous photographic record, through these fog-piercing rays, of whatever lies in front of their bows; while by a method of echo sounding with very short sound waves it is possible to obtain early warning of any near approach to land.

a large number in different parts of the world, apart from those of the rotating-loop type already mentioned. These other beacons are merely ordinary transmitting stations which, in place of programmes, send out regular identification signals to be "direction-found" by ships. The map on page 159 shows the distribution of radio beacons round the British coast. With the "short pulse" system in general use for inter-ship direction-finding, it would clearly be an advantage for these shore-beacons to be converted to the same system. It would then be possible for all the direction-finding required by ships to be carried out on the same wave-length, particular beacons being identified by the characteristic groupings of the pulses they sent out—for example two pulses close together for the North Foreland and two more widely separated for Dungeness.

For ships or aeroplanes which wanted to steer a permanent course on a radio beacon it is even possible to provide automatic lights which will tell the helmsman whether he is keeping to port or starboard of his true course. This is done by an arrangement of buckets or electron traps inside the cathode-ray tube. When the course has once been set, one bucket will catch the electron beam if the ship deviates to port, while a second bucket will come into action if the deviation is to starboard. In each case the arrival of electrons in the bucket produces a small electric current, which is amplified by a valve circuit until it is strong enough to light up a red or green flash-lamp; while a white lamp is lit if the electron beam is centrally placed between the two buckets. By this neat and simple arrangement the last semblance of unfamiliarity is removed from the new method. A red lamp for port, a green lamp for starboard, and a white lamp for straight ahead are symbols the meaning of which is understood the world over.

A slight complication is, however, introduced by currents, whether in sea or air. These are a factor for which no automatic direction-finder can be expected to allow. The course indicated is the true course along the "great circle" which represents the shortest route to the pilot's destination. But if ship or aeroplane is carried to one side by currents it will to that extent go out of its way. Always following the direction signals it will eventually be drawn to the radio beacon, but unless other precautions are taken the course will have been longer than was necessary. This difficulty would not arise if bearings were taken on three or more beacons. Three separate bearings[1] enable the navigating officer to plot his position on a chart with something approaching the accuracy of the stars to which he more naturally turns. Then, knowing two successive positions of the ship and the course steered, he could calculate what the effect of wind and currents had been.

Finally, before parting company with this peculiarly fascinating instrument, we may notice one more example of its extreme versatility. It will be remembered that although the reflected waves are not being used for direction-finding, they have not been eliminated from the records. They are still present as that irregular central "splodge". That means that they could be

[1] Or two radio bearings and a compass bearing.

pressed into service at greater distances, when the ground wave is no longer being received and the reflected wave is more steady. As soon as the reflected wave has ceased to produce irregularities in the receiver, its record will be elongated, like that of the ground wave, into a straight line. So without any alteration to the instrument, it would be possible to use it for long-distance direction-finding. In this way, with powerful beacons in operation on either shore, it should be possible to navigate from the English Channel to New York with complete safety, even though not a glimpse was to be seen of the stars the whole way.

CHAPTER X

RADIO IN WAR

ALTHOUGH no one has any desire to see naval or military radio in active service, modern research has brought about many developments which are of considerable interest, especially to this country. Ultra-short and micro waves in particular have gone far, in their limited field, towards securing secrecy, the first requirement in all war-time communication. England is especially concerned, since, as we shall see later, it looks at present as if the application of ultra-short waves will be most useful at sea and in what are generally described as policing activities.

The whole trouble with radio in war is that it is so painfully apt to prove a double-edged weapon. Not only are broadcast messages liable to be universally "tapped", but the positions of their transmitters can be found by the direction-finding equipment described in the last chapter. These difficulties may be illustrated by a number of stories, taken from experience in the Great War, when the possibilities and limitations of radio were less fully understood than they are to-day.

Both England and Germany established chains of direction-finding stations not only within their own territories but within the zones of military occupation as well. The positions of the two chains of stations are shown in the accompanying map, facing each other as opposing armies are shown in the traditional battle-maps of historians. The Germans used their set of stations, among other purposes, to help the navigation of raiding Zeppelins. They were able to send the Zeppelins radio indications of their position—but

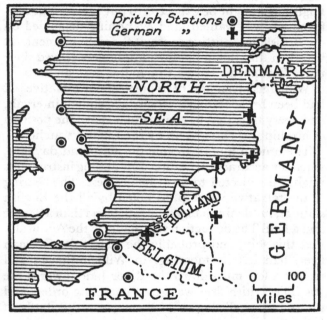

Fig. 21

every time a Zeppelin asked for its position, the message was also picked up and tracked down to its source by the English stations. As a result there was little need of any other form of raid warning, and many a Zeppelin would have escaped destruction or danger if it had been a little less anxious to obtain its position with certainty.[1]

A second example of the indiscreet use of radio paved the way for the Battle of Jutland. At this time the Admiralty already had information that the German High Fleet might be expected to put to sea. But confirmation was wanted, and everything depended on the British fleet putting to sea at the right moment. It was then that the radio direction-finders played their part. One particular warship in the German harbour of Wilhelmshaven had been particularly talkative. It had been broadcasting messages at frequent intervals, and every time that its commander spoke the position of his ship was checked up by the English watchers.

The crucial moment came during broad daylight, when the accuracy of the direction-finding instruments might be expected to be at its greatest. The bearing of the talkative warship, as taken by all the English stations, suddenly changed by rather less than a degree and a half. The deduction was made by the Admiralty that this ship, and probably the rest of the German fleet, had left its moorings in Wilhelmshaven and moved a few miles down the River Jade towards the sea. The British fleet was given its sailing orders and the result, as all the world knows, was the Battle of Jutland.

The English fleet was on the whole more cautious in the use of its radio transmitters. It was realised only too well that our own wireless operators had no diffi-

[1] Since the positions of the Zeppelins were being independently found by the English stations it was also possible to work out the entire code used by the German stations for imparting this information.

culty either in detecting the signals which preceded changes of wave-length on the part of the German transmitters or in locating them when in action. And it was assumed, very properly, that German direction-finders might be equally effective. But a third example may be quoted to show that the mistakes were not all on one side. The scene was the Mediterranean, where England had another chain of direction-finding stations, and the chief actor the enterprising commander of an Austrian submarine.

The trouble in this case arose from the fact that the English stations were in the habit of exchanging by means of wireless messages, the radio bearings which they severally obtained. The submarine commander was perfectly aware of this procedure, and on one occasion when he wanted to know his position took advantage of his knowledge. What he did was deliberately to send out a dummy message. Naturally the direction of the submarine was noted by the English operators, who promptly proceeded to make the necessary comparison of their records through their own transmitters. The submarine commander, so the story runs, waited patiently, was duly told where he was and, after politely acknowledging the information given, went about his business reassured.

The secrecy problem is naturally as much military as naval. It is so acute that most military experts would agree that, from their point of view, the chief use of ordinary broadcasting wave-lengths in war must be for propaganda purposes. Governments can prevent their peoples reading enemy newspapers, but it is difficult, if not impossible, to prevent them listen-

ing to enemy broadcasting stations if they have a mind
to do so. Certainly it would require a very elaborate
and flexible organisation to secure the immediate
"jamming" of all propaganda stations by home transmitters working on the same wave-length, and it is not
difficult to see that the mere giving of perfectly accurate
war news might at times produce an effect which it
would be difficult to counter.

For naval purposes, however, the position is rather
different. Here wireless of wave-lengths long enough
to be reflected from the radio mirrors offers the only
means of long-distance communication which is available; and for the conveying of orders from shore to
ship any means of communication may clearly be
better than none at all, even if the only protection
against eavesdropping is the notoriously fallible one of
transmitting in code. But the ships themselves will
never be anxious to do more talking than is absolutely
necessary. It is true that it has since been admitted
that the Admiralty would never have based their Jutland decision on direction-finding if they had then
known as much as they afterwards learnt about the
possibilities of error with the equipment available.
But the accuracy of modern apparatus would make
promiscuous talking one of the most dangerous pas-
times in which a ship could well indulge. The letters
"d.f'd", which stand for "direction-found", have for
many years now been part and parcel of the naval
officer's vocabulary. They stand also for something
which must, at almost any cost, be avoided.

The nearest approach to a solution of the problem,
so far as ordinary wave-lengths are concerned, is the

invention of a method by which shore transmitters can make reasonably certain that a ship has received a message, although the ship itself has not dared to acknowledge it. The system depends on having a number of different shore stations on either side of the area in which the ship is believed to be steaming. One shore station having broadcast the message, the others will then signal acknowledgment. Then, if replies are received from them all, there will at least be a strong possibility that the ship has also received the message.

If ultra-short and micro waves have not solved all war problems outright, they have at least substantially contributed to the solution of many of them. They have three great advantages. They are limited in range; and micro waves, particularly, can be made strictly directional. Both these are long steps towards secrecy. In addition ultra-short waves are at present comparatively immune from direction-finding by ordinary methods.[1] But it is probable that at least rough direction-finding will, in the long run, be found possible within the range of strong reception; and it is probable that secrecy will be best secured by making the transmitted beam as narrow as possible.

The balance of future advantage would therefore seem to lie definitely with micro waves, which are both more strictly directional and require smaller transmitting and receiving apparatus. It may even be that still shorter radio waves, or perhaps even infra-red waves, may come to play their part in naval communication. Some trials with infra-red waves were in fact

[1] For 5 metre waves, 8 miles has been mentioned as the limiting range for direction-finding with transmitters of the power now available.

made during the last war; and, since they can pierce fog and mist, there seems to be no theoretical reason why they should not be used to carry telegraph and telephone messages. They would be more limited in range than micro waves because they would possess no power of bending round the earth's curvature. But in some ways this would be an asset, and in any case there is no reason why ships should not carry medium-wave, micro and infra-red transmitters, the commander selecting in each case the most suitable method of communication for the particular circumstances. Certainly naval and military experts are both taking a keen interest in ultra-short and micro wave developments, although it would be too much to say that they are in either case convinced adherents of the new cult.

There are at least two possible naval uses. One is for communication within a fleet up to distances of, say, 50 miles. Micro-wave transmitters being both small and easily manipulated it would be a simple matter to turn them in any desired direction. Then, if the position of the ship with which it was desired to talk was roughly known, there should be no great difficulty in establishing communication; while, even if the position of the other ship was not known, it could be found—at a price. A brief and non-committal transmission on a longer wave-length would enable each ship to "d.f." the other. Then, although the enemy fleet would at the same time have had a chance to estimate their positions, they would be able to bring their micro-wave transmitters into action with some assurance of secrecy. No ship outside the direct line of communication would be able to pick up their messages; and it is unlikely in any case that these would be audible farther away than about a hundred, or at the most perhaps two hundred, miles.

This indirect way of getting into touch would obviously not be desirable as a regular procedure. But it would be an advance on anything that has hitherto been available, and there are times when it might well make the difference between success and failure. For example, if two weaker fleets, separated but fairly close to one another, must combine to meet a stronger fleet, then the chance of arranging a secret rendezvous might be worth taking even if it involved giving away their positions at that moment to the enemy. Finally, it may be noted that to secure steady reception in rough weather the micro-wave transmitters could be held steady by gyroscopes, as may already be done with the similar beams of searchlights.

There is one other possible naval use of radio which, although not dependent on ultra-short waves, deserves to be mentioned, if only because of its spectacular interest. It depends on the fact that it is possible to use incoming radio waves to operate the steering mechanism of a ship automatically. This has been done for years with target ships as a regular part of naval training in this and other countries. But in warfare it should mean that operations such as the Zeebrugge raid could be conducted, less romantically, with no risk to personnel. The blockade ship which was to be sunk could both be steered by radio and blown up by radio. The only difficulty would be for the watchers outside to be certain that they were really steering it on the right course, and it is here that the inevitable ultra-short or micro waves may perhaps be introduced.

The first essential would be for the blockade ship to give an accurate indication of its position which the controlling vessel could plot on its own chart. To do this it would be arranged that the blockade ship should regularly broadcast medium-wave signals which could be independently "d.f'd" by two ships outside, preferably stationed some 20 or 30 miles apart. There would have to be two ships, since one direction-finding signal would only give the direction and not the position of the blockade ship. We have now only to put the two watching ships in micro-wave communication and the one which is controlling the blockade ship has all the information that it needs.

In military warfare the chief uses of micro waves would be where either the difficulty of the country or the rapidity of a column's advance made the laying of adequate field telephones impossible. While these conditions would be most often found in the course of frontier expeditions, they might sometimes arise in open warfare generally; and it is just then, when large numbers of units are on the move at the same time, that rapid communication would be most valuable. For this purpose micro-wave transmitters might be almost too directional. It would probably be over-difficult for transmitters and receivers to get in touch. But on rather longer wave-lengths the problem of enabling a large number of transmitters to work at the same time has lately been solved by a particularly ingenious application of the cathode-ray tube direction-finder.

It will be remembered that when a cathode-ray tube

is being used as a direction-finder, the beam of electrons which it produces is drawn out into a straight line indicating the direction of the transmitter. An extension of this idea makes it possible to sort out signals on the same wave-length but arriving from different directions. In order to do this a small movable bucket is provided, inside the cathode-ray tube, which will only catch the electron stream if it has been drawn out in the right direction. With the qualification that secrecy would be to some extent sacrificed it would therefore be possible for a whole series of units, using standard low-powered transmitters all working on the same wave-length, to keep mutually in touch without interfering with one another. The only essential condition would be that no three of the units should ever find themselves in a straight line, when not even the best of cathode-ray tubes could distinguish between signals from their respective transmitters. It may be noticed that the same condition would have to be satisfied by an enemy transmitter which wished to "jam" reception by sending out an independent set of signals on the same wave-length.

Finally, there is one other case in which micro waves would have everything in their favour and no drawback of any kind. We have only to imagine two hill forts, anxious to communicate, but separated by a tribal force or unable to maintain a land telephone link without undue sacrifice of personnel. In the old days they would have signalled to one another, on sunny days, by heliograph. Now exactly the same function could be fulfilled by a pair of micro-wave transmitters. The signals, travelling only in straight lines, could not possibly be picked up by any listener in the valley beneath, and so the risk of leakage would be reduced to an absolute minimum.

Generally speaking, however, radio must be regarded as a supplement rather than as a substitute for the time-honoured field telephone. There are occasions when it is more important to establish communications between friends than dangerous to be overheard by an enemy. On such occasions ultra-short or micro wave radio would be invaluable. In other cases their directional properties could be counted upon to preserve secrecy. It is impossible to say how often these conditions will arise in real life. But for naval purposes, at least, and in frontier warfare, it seems certain that the newest and most mobile of radio transmitters will abundantly prove their usefulness.

CHAPTER XI

RADIO AND THE WEATHER
FORECASTER

I т is not generally realised that the greater success of
the British Meteorological Office in forecasting weather
since the war years is largely due to the increased
number of weather reports received from ships in the
Atlantic. It is English meteorologists, in fact, who
have gained the most in this respect. So much of our
weather comes from across the Atlantic that messages
from liners cannot but be of the greatest value.
Valencia, the Air Ministry's station on the west coast
of Ireland, used to give the first indication of "a de-
pression advancing from the Atlantic". To-day, as our
westernmost fixed station, it is still important but it is
seldom that it gives the first warning of impending
trouble.

Radio is also to a very large extent the basis of the
whole system for the international exchange of weather
information—and there is no other sphere in which the
countries of the world are so mutually dependent. In
Europe and western Russia alone there are more than
six hundred weather-reporting stations. Each one of
these stations sends in its records, usually by telegraph,
to its national headquarters; and it is then that the
process of international exchange begins. All the
countries concerned broadcast their own "synoptic"
reports, containing full weather information from a

sufficient number of stations to cover the whole of their territory. A number of high-powered stations are also utilised to send out what are already international weather summaries, covering a whole group of countries. In addition, the United States Weather Bureau sends out twice a day, similar summaries for its own and Canadian stations, as well as a limited number of reports from the West Indies and the north of South America.

The whole procedure, from start to finish, is so far as possible internationally standardised. Weather observations are made at the same time, and there is an international code in which messages are sent out. Five groups of five figures suffice to convey all the necessary information from each station. The barometer reading uses up three of the figures, while the remainder tell listeners the temperature, wind strength, state of the sky and so on. The weather experts of every country are therefore regularly supplied with an amount of information which, in uncoded form, would take the better part of the day to transmit. It is from these coded messages, quickly deciphered in every meteorological bureau, that the weather maps are built up upon which all forecasts are based; and quite a number of amateurs, using only a single radio receiver, find it possible to make surprisingly complete weather maps for themselves.

The debt of the weather forecaster to radio, and for that matter of the man in the street who uses his forecasts, is thus already very great. There is, however, an entirely new aid to forecasting, based solely on radio observations, which although at present embryonic, may well prove of very great importance in the future. It is based on yet another accomplishment of the cathode-ray tube. As well as improving television and ordinary direction-finding, these remarkably versatile instruments also enable scientists to follow the movements of thunderstorms.

That a cathode-ray tube should have this power is not altogether surprising. Thunderstorms are, as we have already seen, nature's wireless transmitters, and their location is therefore nothing more than a special case of direction-finding with the added difficulty that the signals to be recorded are completely irregular. The method used depends, in the first place, on the same pulling out of the cathode-ray tube's electron beam into a straight line, indicating the source of the atmospheric. It would therefore be possible for an observer, with stop-watch in hand and his eyes continually fixed on the recording screen, to estimate the direction of each separate atmospheric as it came in. But it would also be very laborious, and in practice small cinema films are used instead, which not only save an enormous amount of work but provide a permanent record which is independent of the momentary judgment of the observer.

Thus far we have enabled the watcher of atmospherics to determine the direction from which each is coming. But as with any isolated direction-finding signal there is the ambiguity that the source may lie in either of two opposite directions, that is to say East or West, or North or South; and in any case only the direction and not the position has been observed. The remedy is to keep records in two separate places, in such a way that they can be directly compared and individual atmospherics identified on the two sets of records.

In Britain the two stations used for this purpose are Slough, near London, and Leuchars in Scotland. They are far enough apart to show significantly different directions for even distant atmospherics, and the Post Office supplies the necessary link in the shape of a direct telephone line connecting the two stations. Along this line there passes from Slough to Leuchars, every second, a momentary electrical impulse which, after being appropriately amplified, is impressed on the Leuchars recording film. The same signals are recorded at the Slough end, extra identifiable signals being also provided in both cases at intervals of a minute. To identify any particular atmospheric on the two different records, it is then only necessary for the two observers to note that some particular minute signal is the first after, say, one of the B.B.C. time signals. One correspondence between the time records being known for certain in this way, all subsequent time signals can be accurately identified.

Various tests have shown that the system can be made to operate with surprising accuracy. For example, when a thunderstorm has been proceeding above Slough, the Leuchars observers have been able to identify the direction of particular lightning flashes to within one compass degree. It was even possible, at a distance of 350 miles, to distinguish between flashes taking place a few miles to the west and a few miles to the east of the Slough station. Similarly it has proved possible to obtain the position of atmospherics two thousand or so miles away with very fair accuracy. So that within a few hours of the event weather fore-casters could be supplied with full particulars of all areas within that distance of Britain from which atmospherics were proceeding.

The practical usefulness of this feat depends on the theory of what are called "cold fronts", which was developed by the Norwegian meteorologists, Bjerknes, father and son, during the War when their country was cut off from many of the usual sources of meteorological information. They found that they could do most of their forecasting in terms of the movements of these "cold fronts", a cold front being described as the boundary line where cold polar air was pushing its way under warmer air of southern origin. The usual picture of a cold front shows the polar air on one side of the boundary line, moving to the south-west, while the warmer air, on the opposite side of the boundary line, is moving in exactly the opposite direction. It is easy to imagine that such an arrangement would be un-stable, and mathematical investigations have shown that under those conditions any small disturbance will quickly be magnified. That is the way, according to this theory, in which "depressions" are born.

"Cold fronts" are therefore the home of both thunderstorms and disturbed weather conditions generally. And, while only a certain number of at-mospheric-producing centres can be definitely identi-fied with thunderstorms, it seems probable that all atmospherics originate in areas which, from a meteoro-logical point of view, are disturbed. It is thus possible, purely from radio observations, to trace the position and movements of a cold front. On one historic occasion one of these fronts was traced the whole way across Europe from near the Hebrides to the Black Sea with appreciably more accuracy than could be obtained by the meteorological records available at the time. Only, when six weeks later, records from a larger number of meteorological stations could be compared, was the triumph of the new radio method made clear.

For practical purposes the detailed plotting out of cold fronts from radio records would be most usefully performed by the meteorologists, who would also have the latest batch of weather reports to help them in their task. It is even possible that the cinema films showing the directions of atmospherics might be instantaneously transmitted by television from Slough and Leuchars to the Air Ministry in London. No very great detail would be needed, and it would be possible to send quite good enough pictures by land line. The practical advantages of such a system would be considerable. Each country could at once obtain from its own records a general picture of weather happenings within a whole continent. The weather picture so obtained would, moreover, be continuous and not dependent on reports made only at four stated intervals during the day, as is at present the case.

It is not suggested that atmospherics will ever take the place of the barometer as the mainstay of our weather prophets. For one thing, even the most imaginative of forecasters, except perhaps as far north as Norway, wants more extended information than that provided by the movements of cold fronts. For another, atmospherics seem to be more copiously produced when a cold front is passing over land than over sea. Information from the Atlantic, so often the all-important factor in forecasting, might therefore be deficient if sole reliance was placed on atmospherics. Yet there are occasions when thousands of holiday-makers would like to be given the latest news of a cold front, suspected to be advancing down the British Isles. Nor are the demands of industry for last-minute forecasts likely to become less exacting. A score of interests from paper manufacturers to caterers are concerned, from one point of view or another, with weather conditions; and every year the Air Ministry receives more and more inquiries from commercial and private sources. So whether as a useful second-string at times of regular forecasts, or as a stop-gap in between, it seems inevitable that radio records of at-mospherics will in the long run establish themselves as an important asset in weather forecasting.

ON THE RADIO VALVE

I n all that has gone before, the existence of the wireless valve has been taken for granted. It is the basis of every form of modern receiver, is widely used in transmitters and is essential for the amplification of any electrical signals of any kind. It is equally necessary in broadcasting and in every one of the present and prospective applications of radio which have been discussed in preceding chapters. Outside radio it has an almost infinite range of usefulness.

In a radio receiving set valves are required to do two things:

(1) Acting in the ordinary sense of the word as a valve, to allow an electric current to flow in one direction but not in the other.

(2) To amplify the current produced in the aerial until it is strong enough to operate a loud-speaker.

These two uses of a valve are quite distinct. The necessity for the second is obvious. That for the first requires a word of explanation. It will be remembered (*vide* Chapter Six) that sound is "carried" by radio in the sense that the rapid-vibrating waves of radio are made to trace out, by variations in the extent of their movement, the form of the slower-moving waves of sound. The figure used to illustrate this point is reproduced below, it being only necessary to re-member that the strength and changing direction of the current set up in an aerial exactly correspond with the shape of the radio waves received.

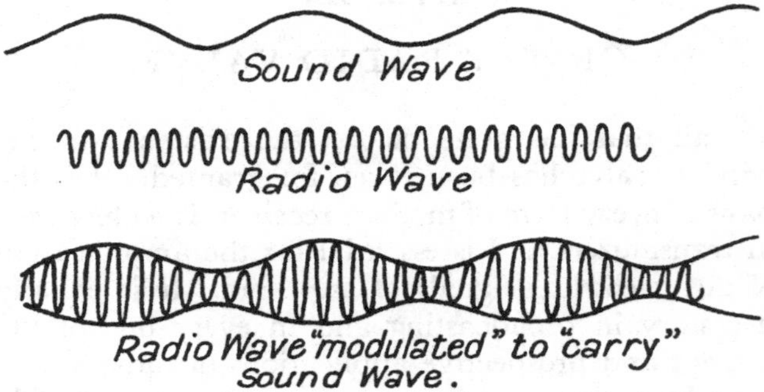

Fig. 22

The final problem is, naturally, to reproduce the sound waves which have been so successfully carried—first through space, and now down our aerial. This implies that something must be set vibrating in the ordinary mechanical sense, since we have seen before that all sound waves are produced by mechanical vibration; and at once the difficulty arises that no mechanical movement, for example that of the parchment cone of our loud-speaker, can be expected to keep pace with the millions of times a second electrical vibrations set up by radio waves. On the other hand the form of the sound waves will be equally well "carried" if only, say, the top half of the "modulated" wave shown in the figure is allowed to reach the loud-speaker. This is precisely the result of only allowing current to flow in the receiving set in one direction. The loud-speaker will still, it is true, receive a larger number of vibrations than it can respond to separately. But, as they are now all in the same direction, a whole sequence of rapid radio vibrations can now be used to build up the mechanical movement required to produce a single sound wave.

The production of this "one way" effect by a radio valve depends in the first instance on the fact that the flow of electricity along a wire means the transfer from one end of the wire to the other of particles of so-called "negative"[1] electricity, that is, electrons. In the ordinary course of events these particles can flow in either direction. But it so happens that any metal, when heated, has a capacity for throwing out electrons from its surface. Part of the metal is, literally, boiled off; and, at any rate in the case of some metals, the part that is boiled off consists almost entirely of electrons. Here, therefore, at once is the possibility of a one-way flow of electricity. If an electric current is to flow across the gap between a hot wire and a cold wire, it can only be in one direction. The hot wire can throw out electrons to be caught by the cold wire. But the cold wire cannot reciprocate.

In practice the cold wire is replaced by a metal cylinder placed round the hot wire, so that as many electrons as possible can be caught. The cylinder is also charged positively, by means of a battery, so that the negative electrons are attracted to it. Finally both the hot wire and the cold cylinder are encased in a vacuum bulb so that the electrons' journey from

[1] The term "negative" is purely conventional in origin. Since it turned out that the natural unit of practical electricity was the electron, it would have been more logical if the terms positive and negative had been inverted.

the one to the other is made as easy as possible. That is the diode valve, so called because it has only two essential components, the hot wire and the cylinder. The one-way possibilities of such an arrangement were first specifically pointed out by the American inventor, T. A. Edison, in 1883, although he was not able to give any physical explanation of how they arose. About the same time the English scientist, Sir Ambrose Fleming, published the results of some experiments on similar lines and some years later gave the correct explanation in terms of the throwing off of electrically charged particles—"electrons" as such not having yet been discovered. Finally, when Sir Ambrose Fleming became associated with Marchese Marconi in the design of the Poldhu Transatlantic station in 1899, he saw the possibilities of such an arrangement in wireless reception—and five years later he made the first radio valves. This was in 1904, two years earlier, it is interesting to notice, than the invention of the first of the now almost defunct crystal detectors.

The diode valve provided a means of detecting radio waves, but it could not amplify the signals received. For amplification the triode valve was needed, the invention of the American, Lee de Forest, in 1907. As its name suggests, the triode is a diode with something added. But although it is in a sense the child of the original Fleming valve, the something added is so important that it can almost be regarded as a separate invention.

What Lee de Forest did was to interpose what is called a "grid" between the hot wire and the cold cylinder. The grid consists of a zig-zag wire and its action may be compared with that of traffic lights. If the electrical pressure applied to the grid is such that electrons are repelled by it, i.e. if it is negatively charged relative to the hot wire, then the control light is set at "Stop". No electrons can get past the grid to the metal cylinder beyond. If, on the other hand, the grid is positively charged, it attracts the electrons thrown off by the hot wire. But as the grid consists chiefly of holes the greater part of the electrons rush rapidly past it and arrive safely at the metal cylinder. The control light is then at "Go". Fortunately, quite a small difference in electrical pressure is enough to make a big difference in the flow of electricity from hot wire to cold cylinder. In this way the small electrical changes produced in the original circuit of a radio receiver can be magnified and, by using a succession of valves, the process can be repeated more or less indefinitely. Although there are many more complicated types of valve, ranging up to the octode which has eight different components instead of only three, they can all be broadly described as arrangements for making one valve do the work of two or more valves.

As well as providing amplification in a receiving set, the triode valve is also used at transmitting stations as a source of electrical oscillations. How this is done is a more complicated story, into which it is not proposed to enter fully here. It is sufficient to say that it is possible to link up the grid with the hot wire and cold cylinder circuit in such a way that the combined arrangement is unstable. This means that any electrical disturbance, however small, will automatically build itself up, an effect which it was formerly possible to illustrate, to the annoyance of Post Office engineers, with an old-fashioned telephone. By holding the earpiece up to the mouthpiece, the energy released by any movement of the microphone could be " fed back " until a sufficient disturbance had been built up to produce a continuous and melancholy howl. The result of a similar " building up " process, in the case of the triode valve, is a rapid to and fro movement of electricity which can be conveniently used as a source of radio waves. So that even if valves lie outside the main story of radio waves, as told in this book, their importance, both to wireless engineers and builders of receiving sets, is wellnigh universal.

INDEX

INDEX

Printed in the United States
By Bookmasters